Environmental
Design
Best Selection
4

快適生活時代のキーワードは「環境」と「デザイン」

日本ではいま，環境問題が毎日のようにマスコミをにぎわしている。かけがえのない地球を守るためにどうすべきか，議論が進み，具体的提案も多くなっている。環境保全のための努力が，さまざまな分野で展開されつつある。

その努力はもともと，政府をはじめ公共団体による行政レベルはもちろんのこと，企業レベル，さらには生活者レベルで，それぞれに国際的なスケールで行われなければならないものである。なかでも企業の，環境問題へのチャレンジが最も実質的で，また大きな影響力をもっている。そのあたりをもう少し具体的にいえば，企業ワークの3分野，つまり〈1〉テクノロジー（技術），〈2〉プロダクション（生産），〈3〉マーケティング（対市場活動）のそれぞれのなかで，もうすでに「環境」をテーマに，活動しはじめているのである。

スウェーデンの自動車メーカー，ボルボ社は昨年，日本の新聞に「私たちの製品は，公害と，騒音と廃棄物を生みだしています」というショッキングな見出しのあとに「だからこそボルボは，環境問題に真剣に取り組みます」と宣言，そのための改善をアピールした。実に大胆な「環境マーケティング」の実例である。日本の企業もまた，避けて通れない重要課題が「環境」である。

また一方で，地球レベルで「境界のない時代」が進んでいる。西欧と東欧の対立がやわらぎ，まさかと思われたベルリンの壁までが取り払われた。男性と女性も社会的境目がなくなり，仕事と遊びを融合して生活する人間が増えた。同じようにデザインの世界でも，製品デザインと環境デザインといった，従来のジャンル分けが意味をもたない状況にある。このようにすべての面で，本格的なボーダレスの時代を迎えている。こ

高層化と高密度テクノロジーによって，日本の都市空間のイメージは大きく変わった。公共空間，商業空間，オフィス空間，住空間なども，リッチなニュー・イメージへ展開中である。また，開発系では，リゾート計画，ウォーターフロント計画が本格化しはじめている。1990年代の半ばには，環境デザインの主流になるような大きなプロジェクトが日本列島を覆いつくす勢いである。

「環境」と「デザイン」は上に述べた理由で，1990年代の重要なキーワードになるに違いない。そうした状況のもとで，この環境デザインの秀作を集めた「環境デザイン・ベストセレクション」を出版できたことを，心から喜びたい。そしてこの作品がそれぞれに発信している内容は，現代の事実であり，近未来の予言である。これらの情報を真剣に受け止めることによって，未来のデザインワークに好ましい影響を与えるに違いない。

「環境デザイン ベストセレクション 4」編集委員会 寺澤勉

れまでの対立や区別の概念は，融合・拡大へ転換している。そうした意味では，これまでのイデオロギーは「再編成と統合化」の時を迎えている。そうした意味で，境界を前提にしない「環境デザイン」の概念が時代の中心を支える，そんな趨勢が見える。

日本人の生活の豊かさを実際に展開し，生活の快適さに深く迫り，実現する手段として，環境デザインの発想とテクノロジーが生きている。製品のデザインでも，個々のモノの機能と形態だけを考えていては，よいデザインはできない。人と人を取りまく環境，グローバルな視点と地球をも意識した発想が，すべてのデザインに必要である。いま日本の環境デザインは，エキサイティングで，魅力的である。環境デザインの状況はまさに「ニュー・デザインイヤー」，異常なほど盛りあがっているのは，こうした状況を人々が求めているからである。

Keywords for the age of amenities are ENVIRONMENT and DESIGN.

The environmental issue is currently being discussed on a daily basis in Japan. Many arguments have been exchanged over the issue of exactly how to go about saving our irreplacible earth, and many concrete proposals have been put forward.

Various acts of preserving the environment can be observed in many fields. This needs to be approached not only at the level of administration for public bodies, including government, but also at the level of consumers on an international scale. Above all, the challenge on the corporate level is the most substantial and effective. Allow me to offer a concrete example of this. The three fields at the corporate level; <1> technology, <2> production, and <3> marketing, have already started action based on the theme of ENVIRONMENT.

Volvo, a Swedish auto maker, came up with a shocking headline in a Japanese newspaper. It stated, Our cars create environmental pollution, noise and waste!. An added-on statement said, That's why Volvo tackles enviromental issues very seriously, and thus appealed for improvement. This is an example of bold marketing. The one important issue which Japanese companies cannot avoid is the environmental issue.

On the other hand, THE BOUNDLESS AGE is in progress at the level of earth. The confrontations between east and west Europe have been taken to task, and the Berlin Wall, which nobody ever expected to see the end of, came tumbling down. Social boundaries between men and women have disappeared and many people now find it easier to marry their work and leisure together. The same can be said for design. The former genre, productive design and environmental design no longer have any meaning. Thus, we are all experiencing a full-scale era of reduced borders in all respects. Former concepts such as confrontation and distraction have been converted to harmony and extention. In this sense, ideology up until now is facing a period of reorganization and integration. This means that we appear to be heading towards an age that is supported by the idea of boundless enviromental design.

The conception of environmental design and technology is creating the actual development of wealthy lives for the Japanese people and seriously approaching the materialization of a life abundant in amenities. Even when it comes to the design of a product, if one thinks only of the individual function and shape necessary, a well designed product cannot be produced.

All designs should involve the environment concerned, the global view and the concept of global awareness. Very exciting and attractive environmental designs are available at the moment in Japan. It has been a very prosperous year for environmental design, and has been named THE YEAR OF NEW DESIGN. The reason for this extraordinary boom in environmental design is due to the fact that the consumers are demanding it.

High-rise buildings and high density technology have greatly changed the image of Japanese city space. Such areas as public space, industrial space, office space and living space are in the process of changing their image to something new and rich. Additionally, as far as development is concerned, full-scale projects for resorts and waterfronts have commenced. By the middle of the 1990's, the whole of Japan will be covered with large projects based on environmental design if the boom continues as it is.

Environment and Design will certainly become important keywords for the decade of the 90's owing to the reasons mentioned above. Iam very happy to be able to announce the publication of this book called A Best Selection of Environmental Design, a collection of excellent work based on environmental design. The message stated by each work is connected to the modern age and also to the production of the near future. The future of design work is greatly dependent on how seriously this information is taken in.

Editors of Environmental Design Best Selection 4 : Tsutomu Terazawa

目次

まえがき ·· 2
快適生活時代のキーワードは「環境」と「デザイン」

1. 地域開発・ウォーターフロント計画 ·· 9
エキサイティングな水際へのアプローチ
〈天保山ハーバービレッジ〉 環境開発研究所 杉本正顕
広域都市計画・水辺(湾岸, 港, 河川, 湖沼)計画・地域開発・これらの外構・造園デザイン・他

2. レジャー環境・大型複合商業施設開発計画 ····················· 23
「光」と「水」と「緑」がマイルドに融合したリフレッシュ・ゾーン
〈リーガロイヤルホテル新居浜, リーガアクアガーデン新居浜〉 日建設計
各種レジャー・スポーツ施設・ホテル・ショッピングモール・デパート・テーマパーク・これらの外構・造園デザイン・他

3. 公共・オフィス空間デザイン ·· 87
地域と共存する次世代のインテリジェント・オフィスビル
〈NEC スーパータワー〉 日本電気株式会社
福祉施設・医療施設・交通施設・庁舎・銀行・学校・ビジネスビル・これらの外構および造園デザイン・他

4. ランドスケープデザイン ··· 163
地域社会との融合をめざす創造拠点
〈かながわサイエンスパーク〉 株式会社日本設計
都市計画・タウンスケープ・道路計画・公園計画・広場計画・博物館・これらの外構・造園デザイン・他

5. 住環境デザイン ··· 185
スペースとパターンが織りなす住空間ハーモニー
〈CITY SCREEN XIII 岸田邸〉 池上俊郎＋アーバンガウス研究所
独立住宅・集合住宅・セカンドハウス・これらの外構および造園デザイン・他

6. サイン/ストリートファニチャー・デザイン ······················· 211
サインのフォルムに近未来志向を見た
〈スペースワールドサイン・環境計画〉 株式会社丹青社サインデザイン研究所
街区サイン・案内サイン・交通サイン・舗装路面計画・モニュメント・照明塔・ベンチ・灰皿・屑篭・水のみ・電話ボックス・バス・ストップ・各種シェルター・他

7. その他 ··· 235
神社, 仏閣・史跡保存計画・基地計画・これらの外構・造園デザイン・他

索引 ·· 249
作品名, ディレクター, デザイナー, 撮影者, 応募代表者住所・連絡先

あとがき ·· 262

CONTENTS

Forword ··· 4

Keywords for the age of amenities are ENVIRONMENT and DESIGN.

1. Development of Regions-Waterfront Project ····································· 9

An Exciting Approach to Waterfronts.

〈Tenpozan Harbor Village (Tenpozan Market Place)〉 Environmental Development Research, Masaaki Sugimoto

City plannings of a large region, Waterfront plannings, Development of regions, Related exterior design and landscapings, etc.

2. Commercial and Leisure Environmental Design ···························· 23

Harmonizing refresh zone with SUNSHINE, WATER and GREENERY

〈RIHGA Royal Hotel, Niihama & RIHGA Aqua Garden, Niihama〉 Nikken Sekkei Ltd

Leisure Facilities, Sports Facilities, Hotels, Shopping Malls, Department Stores, Theme Parks, Related exterior design and landscapings, etc.

3. Public & Office Environmental Design ·· 87

An intelligent office building for the next generation that will coexist with the local environment.

〈NEC SUPER-TOWER〉 NEC Corporation

Welfare and social facilities, Clinics, Transport facilities, Municipal offices, Banks, Schools, Business buildings, Related exterior design and landscapings, etc.

4. Landscape Design ··· 163

A Creative Base point Aimed at Integration with the local Community

〈The Kanagawa Science Park〉 Nihon Sekkei Ltd.

City plannings, Townscapes, Road plans, Park plans, Plazas, Museums, Related exterior design and landscapings, etc.

5. Residential Environmental Design ··· 185

Harmony of interior space with pavement pattern

〈CITY SCREEN XIII KISHIDA R.〉 Toshiroh Ikegami + Urban Gauss

Independent houses, Duplex houses, Holiday homes, Related exterior design and landscapings, etc.

6. Sign, Street Furniture Design ··· 211

Expectation to the near-future is found in the forms of sign.

〈Sign & Visual Environment of Space World〉 Tanseisha Sign Design Institute

Street signs, Guide signs, Traffic signs, Pavement and sidewalk plannings, Monuments, Streetlamps, Benches, Ashtrays, Trashcans, Drinking fountains, Telephone boothes, Bus stops, Shelters, etc.

7. Others ··· 235

Shrines and temples, Preservation of historic relics, Related exterior design and landscapings, etc.

Index ··· 249

Project name, Director, Designer, Photographer, Representatives Contributors Address & Telephone, etc.

Postscript ··· 262

凡例（データの読み方）

a：作品名
b：所在地
c：クライアント
d：ディレクター/プランナー　等
e：デザイナー/エージェンシー　等
f：施工者/製作者　等
g：撮影者
h：コメント
i：主な材料
j：応募代表者

LEGEND (A Guide to How to Read Data)

a : Project name
b : Site
c : Client
d : Director/Planner etc.
e : Designer/Agency etc.
f : Contractor/Production etc.
g : Photographer
h : Comments
i : Principal Materials
j : Applicant

Environmental Design Best Selection 4

Copyright © 1991 by Graphic-sha Publishing Co., Ltd.
1-9-12 Kudan-kita, Chiyoda-ku, Tokyo 102, Japan
ISBN4-7661-0639-3

Printed in Japan
First Printing, Jun. 1991

Chapter 1

地域開発・ウォーターフロント計画
Development of Regions-Waterfront Project

エキサイティングな水際へのアプローチ　天保山ハーバービレッジ

環境開発研究所　杉本正顕

⟨1⟩大阪のウォーターフロント開発のパイオニア

これまで都市のウォーターフロントは巨大な産業装置により占有され，都市生活者から完全に忘れ去られた存在でしかなかった。

いま，産業構造の再編成によるウォーターフロントの再生，活性化が取りざたされるに至り，再び都市機能の重要な役まわりがめぐってきたといえよう。

その昔，港はモノ，ヒト，カネの流入，流出の窓口であった所であり，多くの場合現在の都心から至便の所に位置し，非常にポテンシャルの高い立地条件を有している。

一方，高ストレスの都会人，ビジネスマンに水辺を開放し，本物の自然（荒れた海，やさしい海，陽光あふれる水面，そよぐ風，赤い夕日，潮のにおいなど）を取り戻し，うるおいのある都市空間を創出することは，いま都市計画上最重要課題である。

まさに天保山ハーバービレッジ計画は，大阪市における真の意味でのウォーターフロント開発のリーディングプロジェクトとして位置づけられる。

<2>天保山ハーバービレッジの概要

大阪市南区の天保山外航客船ターミナルに隣接する倉庫群の隣地
4ヘクタールに，世界最大級の水族館「海遊館」を核施設とし，レストラン，
ナイトクラブ，物販店からなる「マーケットプレース」でにぎやかなお祭りの雰囲気
を盛り上げ，それらを取りまくプラザ，サンセットプラザ，ディナークルーズ船，
副都心（オオサカ・ビジネス・パーク）との水上アクセスとなる水上バス，ヨット，
モーターボートの展示施設など多彩な諸施設が人間性豊かなアーバン
リゾートを構成している。

<3>海遊館を設計して
　　ピーター・シャマイエフ（approach '90秋号より抜粋）

日本から帰る飛行機の中で二つのアイデアが確立された。
一つは太平洋全体を主題として扱うことである。太平洋の中央部とともに
その周辺部の特徴的な海洋環境を展示し，それにより太平洋を一つの総合体
として，あるいは全体的なシステムとして表現するのである。
二つめのアイデアは建物の内部の重層化である。水槽とエアスペース
（展示通路）を交互に配置することにより来館者を海中に導き，来館者は目の前
の水槽を通して他の海洋環境や海中のエア・スペースを見渡すことができる。
こうした設計により，劇的効果や神秘的な雰囲気，感情的インパクトをもたらす
と同時に，人間を含めた生命あるすべての生物の相互の結びつき，
というテーマを表現できるかも知れない。
「リング・オブ・ファイア」は科学者たちが太平洋に与えている名称である。
太平洋の地殻構造プレートの縁には環状の火山帯が連なっていることから
名づけられた。私たちはそのことを知り，その名前を新しい水族館の名称として，
またメタファーとして確立させた。また火山帯は太平洋沿岸地域を形成し，
そこには豊かな，そして多様な生命体と生息環境が存在していることも知った。
地球の原初的なプロセスである火山活動から出現した生命体と生息環境を
讃えること，それが私たちのテーマである。そして壊れやすい地球環境に対する
人間としての理解を高めること，それが私たちの使命であろう。
太平洋を一周する旅は海面上から始まる。地上の環境を観察し，それから
海の中へと進み，多様な海の生息環境と生命体の神秘，驚異，不思議を
体験する。私たちはこの旅が魅力的で価値あるものになることを望んでいる。
さらに望むことは，この日米の協同の成果であるプロジェクトが二国間のより
大規模な協同を可能にする契機となることである。すなわち次の数十年間に亘り，
人類およびすべての生命体に代わって地球全体を保護，管理するという重要な
責務を担うために，国際間の緊密な協同が望まれるのである。

<4>日本初のフェスティバル・マーケットプレース

ボルチモアのハーバープレース，ボストンのファニェルホール・マーケット
プレース，ニューヨークのフルトンマーケット，ピアー17など全米で70以上もの
大型プロジェクトを手がけ「都市の魔術師」と称されるジェームス W.ラウス氏が
当プロジェクトに参画し，氏の提唱する「フェスティバル・マーケットプレース」を
日本で初めて完成させた。
水辺を開放し，自然を取り戻し，再び人の集まるにぎわいのあるウォーター
フロントに再生する際に祝祭空間をイメージの中心に置いた開発手法であり，
アメリカンテイストそのままで予想以上の大ヒットを続けているのは，「海遊館」との
相乗効果に頼るところは大とはいえ，都市生活者の希求する洋の東西を
問わない普遍的な手法であることの裏返しであろう。
「フェスティバル・マーケットプレースはショッピングセンターと違って非常に
大きな人口規模が必要で，立地は交通至便で人を引きつける魅力があり，
目立つものではなくてはならない。そうでなければささえきれないものだ」
J.W.ラウス氏の言葉である。
「天保山ハーバービレッジは年に1,000万人くらいの集客効果を生むもの
であると思います。成功するためにはそれくらいのものでなければなりません。」と
これまでの実績と背景に自信たっぷりに語る姿には，少しもたかぶったところがなく，
たんたんとした語り口であった。

<5>今後の課題

1). 水族館はすぐれて装置産業であるので，3年，5年，10年という節目ごとに
　　追加投資が必要であり，心づもりはいまから必要。
　　シアター，ミュージアム，ホール，展示場。相乗効果の期待しうるものが
　　あげられる。
2). ホテル，コンドミニアムなど夜間人口が周辺に配置できたとき，
　　真の意味の都市開発が完結したことになろう。
3). パラディッソ，北港マリーナ周辺，スポーツアイランド構想などとの有機的な
　　関連づけと連携を図り，点から面へ，面からネットワークの形成へと広げていく
　　ことが大阪の湾港開発の期待されているところであろう。
　　J.W.ラウス氏がいみじくも語っていた言葉を思い出す。
　　すなわち「ウォーターフロント再開発の最大の受益者は地主である。
　　これに続く，2期，3期へと開発をつなげていくことが最も大事な点である」と。

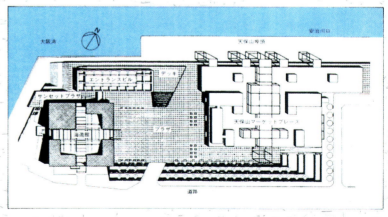

An Exciting Approach to Waterfronts.
Tenpozan Harbor Village (Tenpozan Market Place)

Environmental Development Research, Masaaki Sugimoto

⟨1⟩ A waterfront development pioneer in Osaka.

In the past, urban waterfront areas were occupied by huge industrial machinery and were completely overlooked by the people who lived in the cities.

Now that the subject of reproduction and activation of waterfront areas has once again arisen due to the reorganization of industrial construction, it appears that the time is ripe for these waterfronts to yet again play an important functional role in the city.

A long time ago, most harbors were located in the most convenient areas to act out their liason role for the incomings and outgoings, not to mention the cash flow, of all major cities. These locations were all in highly favourable and potential areas.

On the other hand, one of the most important themes for current city planning is to open up these waterfront areas for the use of the urbanites and businessmen – who suffer from increasingly higher bouts of stress – and bring in a whiff of real nature (rough seas, gentle seas, the surface of the water bathed in rich sunlight, breezes, the red evening sun, the smells of the sea water, etc.) in order to create charming urban spaces.

The Tenpozan Harbor Village is the very leading waterfront project in the city of Osaka.

⟨2⟩ The outlines of the Tenpozan Harbor Village.

The whole complex occupies 4 hectares of a site on which used to stand the warehouses of the Tempozan passenger liner terminal in Minami-ku, Osaka. It boasts various facilities such as the world's largest scaled aquarium, the core of the whole facility, which stands in one corner, the "Market place" which consists of restaurants, night clubs and shops and in which a festive atmosphere reigns supreme in the sunset plaza that surrounds it. There are also dinner cruises, water buses which give access to the heart of the city, and a yacht and motor boat exhibition, all of which go to form an urban resort rich in humanity.

⟨3⟩ Peter Chermayeff, the designer of the Aquarium.
(Taken from the fallissue of Approach).

While flying back from Japan, two ideas took hold. One was that we should address the entire Pacifific Ocean as our subject, as a unity, or total system, by selecting a small number of representative habitats around the ocean perimeter to exhibit, together with the central ocean itself. The second idea was that we could immerse human visitors in the presentaion, through layering, through an alternating sequence of water volumes/air spaces/water volumes/air spaces, so that observers might see through from one marine environment to another, and see other people suspended in the air spaces between. Here might be a strong potential for drama, for mystery, for emotional impact, and for a theme of interconnectedness, of unity among all living creatures, including man.

The "Ring of Fire" took hold as our title, and metaphor, as we learned that scientists give that name to the Pacific Ocean, because of the ring of volcanoes at the edges of the tectonic plates. We found that the ring of volcanoes corresponds quite closely to the Pacific Rim, where habitats and life forms are also most rich, and most diverse. To celebrate habitats and life, emerging from the fires, from the primal processes of the planet – that would be our theme. And to increase human understanding of the planet's fragility, even at the scale of the vast Pacific Ocean – that would be our mission.

We hope that the visitor's journey, around the Pacific Rim, first above the water, observing terrestrial settings, and then underwater, experiencing the mysteries, wonders, and surprises of marine places and animals, will prove to be a rewarding experience, even when the Aquarium is crowded. And we hope above all, that this project, a collaboration of US and Japanese teams, may serve as a useful indication that our two nations are capable of collaborating on a much larger scale, undertaking, in the next decades, conservation and stewardship of the entire planet on behalf of all humanity and all life.

⟨4⟩ Japan's first festival Market Place.

James W. Rouse, known as the city magician, has worked on more than 70 projects throughout the whole of America such as the Harbor Place in Baltimore, the Festival Hall Market Place in Boston, and the Fulton Market and Pier 17 in New York. Having joined this project he completed his proposal for Japan's first "Festival Market Place". The water's edge of a waterfront is opened to the citizens in order to bring back nature and create a zone in which people can getaquainted. This was the method of placing the festival space in the center of an image. The reason for the continuation of the unexpected smash hit that it produced in Japan – well in keeping with its fondness for American tastes – can be largely laid at the door of the involvement of the Aquariumas well as the universal theme of urbanites who maintain a deep desire for the sea.

"The Festival Market Place requires a large scale area which can accomodate a great many more people in comparison to a shopping center. It must be located within easy access, be able to attract many people and stand out. Otherwise it will be impossible to maintain", said Mr. Rouse. "I think the Tenpozan Harbor Village will attract 10, 000,000 people a year, and, in fact, it must reach this figure in order to succeed". He said this in a voice full of confidence, and going by his past record, one can presume that he was not at all haughty, but philosophical.

⟨5⟩ The future theme.

1). As the Aquarium is proving to be an excellent industry device, additional funds will be required in its 3rd, 5th and 10th year. It is necessary to bear this in mind now. Such additional attractions as a theater, a museum, a hall and an exhibition area have the nominations at present.

2). When the surrounding area is able to attract people at night, the true meaning of a city development will come into its own by the construction of such amenities as hotels and condominiums.

3). The development of the formation from point to aspect and from aspect to network through the organic planning connections and cooperation of such places as Paradiso, the Hokuko museum and its surroundings, one can expect that the idea of a sports island will be brought to the waterfront development in Osaka.

I recall the vigorous words of Mr. Rouse, "The person who gets the most profit in the redevelopment of waterfront areas is the landowner. The most important thing is to continue the investment into the second and third development stages after this."

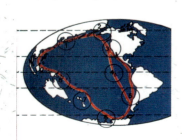
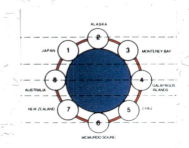
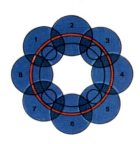
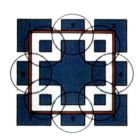

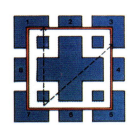

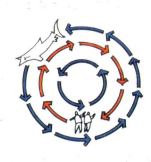

バナーで彩られたマーケットプレース。
Multi-colored flags bedech
the expansive forecourt.

高さ約50m，8階建て，鮮やかな色彩の世界最大級の海洋展示施設，海遊館。建物基部のブルータイルは海，上部の赤いガラスは火山，頂上部の透明ガラスは生命を象徴している。

Blue tile walls with a motif by Ivan Chermayeff suggest the ocean inside, red glass hints at volcanic fire, and the crystalline glass roof symbolizes terrestrial life.

名称＝天保山ハーバービレッジ
所在地＝大阪市港区海岸通1-3
発注者＝大阪ウォーターフロント開発
設計者＝ケンブリッジ・セブン・アソシエーツ，環境開発
研究所
マーケットプレース＝エンタープライズディベロブメント
社（商業開発コンサルタント）
施工者＝竹中・大林・鴻池 J.V.
工費＝215億円（総事業費）

〈建築概要〉
敷地面積＝3万4713平方メートル
建築面積＝1万8938平方メートル（駐車場含まず）
延べ面積＝6万3024平方メートル（駐車場含まず）
構造・階数＝RC造・一部S造，地上8階（海遊館）。RC
造，地上4階（サポートビル）。S造，地上5階（マーケ
ットプレース）

◆海遊館（西ブロック）データ
〈設計協力〉
展示計画＝LYONS/ZAREMBA INC.
展示設計（水槽内）＝THE LARSON COMPANY
設備＝JOHN ALTIERI CONSULTING ENGINEERS＋環
境開発研究所
構造＝WEIDLINGER ASSOCIATES＋環境開発研究所
水処理設備（L.S.S.）＝ENARTEC CONSULTING ENGI-
NEERS＋INTELLICON＋環境開発研究所
グラフィック＝CHERMAYEFF & GEISMAR＋スタジオイッ
クス＋キッチン＋環境開発研究所
展示照明コンサルタント＝HAWARD BRANDSTON
LIGHTING
AVコンサルタント＝FREO BRINKS COMPANY
サウンド＝SOUND DESIGN STUDIO
造園設計＝小山鐵夫＋西 良祐

◆天保山マーケットプレース（東ブロック）データ
〈設計協力〉
グラフィック＝CHERMAYEFF & GEISMAR＋環境開発研
究所

◆海遊館仕上材料
屋根＝アスファルト断熱防水コンクリート押え
外壁＝特注色小ロタイル貼り（イナックス）一部壁画貼
り，コンクリート打放し化粧目地撥水剤塗布
開口部＝特殊複層ガラス（旭硝子），赤色ガラス（P.P.G.）
サッシ＝アルミフッ素樹脂塗装（神戸製鋼，不二サッシ
ュ，三共アルミ）
人工地盤＝コンクリート直押えアスファルト防水コンクリ
ート押えインターロッキングブロック

〈展示通路・エントランスホール・オーディトリアム〉
床＝タイルカーペット重歩行用（長谷虎紡績）
壁＝コンクリート打放し下地調整の上にEP塗装
天井＝PBt9下地岩綿吸音板t12貼りEP塗装

〈5階ラウンジ，海遊館〉
床＝大理石貼り
壁＝ナラ材SOP拭取り仕上
天井＝PBt9下地岩綿吸音板t12貼りEP塗装

◆天保山マーケットプレース 仕上材料
屋根＝カラー折板t0.6断熱材岩綿吹付けt30耐火被
覆t10，アスファルト防水コンクリート押えt60，アスフ
ァルト露出防水屋根スラブ裏断熱岩綿吹付けt30
外壁＝成形繊維セメントt30（アスロック，ノザワ）スチー
ルパネルマリンペイント2回塗り
開口部＝透明フロートガラス，透明網入りガラス，型板
網入ガラスt6.8
サッシ＝アルミアクリル焼付けスチールマリンペイント2
回塗り

〈パブリック通路〉
床＝エポキシ系塗り床，一部磁器タイル貼り100角
壁＝PBt12貼りEP塗装
天井＝スチールメッシュ150×150グリッドSOP塗装

◆撮影＝吉村行雄

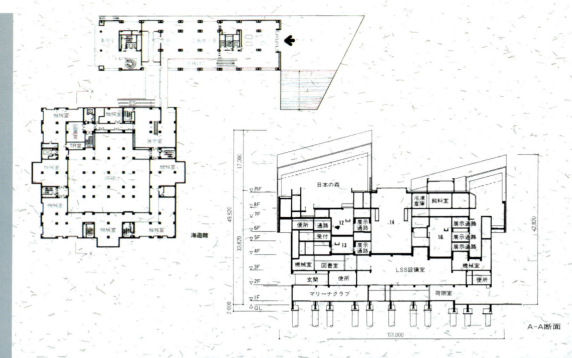

日本の森

A-A断面

海遊館

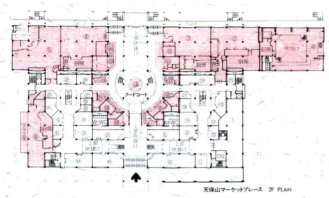

天保山マーケットプレース　2F PLAN

Project title=Tenposan Harbor Village
Site=1-3, Kaigan Dori, Minato-ku, Osaka
Client=Osaka Waterfront Development
Designer=Cambridge Seven Associates, Envi-
ronment Development Research
Market place=The Enterprise Development
Company (Commercial Development Consul-
tant)
Contractor=Takenaka, Obayashi, Konoike J.
V.
Construction costs=¥215,000,000,000
(project total)

〈Design Outline〉
Site area=34,713 square meters
Building area=18,938 square meters (ex-
cluding car parking space)
Total floor area=63,024 square meters (ex-
cluding car parking space)
Construction/floors=RC construction, part S
construction, 8 stories above ground (aquarium). RC construction, 4 stories above
ground (support building). S construction, 5
stories above ground (market place).

◆ Aquarium (West block) data
〈Design cooperation〉
Display plan=Lyons Zaremba Inc.
Display design (in cistern)=The Larson Co.
Equipment=John Altieri Consulting Engi-
neers + Environment Development Research.
Construction=Weidlinger Associates + Envi-
ronment Development Research.
Waterworks(L.S.S)=Enartec Consulting Engi-
neers + Intellicon + Environment Develop-
ment Research.
Graphics=Chermayeff & Geismar + Studio
Ikks + Environment Development Research.
Display lighting consultant=Haward Brand-
ston Lighting.
AV consultant=Fred Brinks Company.
Sound=Sound Design Studio.
Garden design=Tetsuo Koyama + Yoshihiro
Nishi.

◆ Tenpozan Market Place (East block) data
〈Design cooperation〉
Graphics=Chermayeff & Geismar + Environ-
ment Development Research.

◆ Aquarium Finishing materials
Roof=Asphalt insulating with waterproofed
concrete pressed
External walls=Special order color tiles
(Innax), part fresco tiling, exposed concrete
simulated joint waterproofing.
Opening=Special order double glazed glass
(Asahi Glass), red glass (P.P.G.). Sash=
Aluminium flourine resin painted (Kobe Steel,
Fuji Sash Co.,Ltd, Sankyo Aluminium Co.,
Ltd).
Artificial foundations=Concrete direct support
asphalt waterproof concrete press interlocking
blocks.

〈Display alley, entrance hall, auditorium〉
Floor=Tile carpeting with reinforcing treat-
ment for walking(Hasetora Weaving Co.,
Ltd).
Walls=EP paint on ground adjusted exposed
concrete.
Ceiling=PBt9 asbestos soundproofing,
ground slate, t12 EP panelling.

〈5th floor lounge, Aquarium〉
Floor=Marble flooring
Walls=Oak SOP wipe finish
Ceiling=PBt9 asbestos soundproofing,
ground slate, t12 EP panelling, EP coating.

◆ Tenpozan Market Place, Finishing materials
Roof=Color folded slate 0.6, insulate asbestos spray t30, fireproofing t10, asphalt waterproofing concrete press t60, exposed asphalt
waterproofing roof slab lining, insulate asbestos spray t30.
External walls=Plastic fibre cement t30 (As-
lock/Nozawa), double coated steel panel
marine paint.
Opening=Transparent float glass, reinforced
transparent glass, reinforced figured glass t6.
8. Sash/Aluminium acryl enamelled steel,
double coated marine paint.

〈Public walking spaces〉
Floor=Epoxy painted floor, 100 square part-
ceramic tile flooring.
Walls=PBt12 panelling EP coating.
Ceiling=150×150 steel mesh grid with SPP
coating.

◆ Photographed by Yukio Yoshimura

a：三井倉庫箱崎ビル
b：東京都中央区
c：三井倉庫オフィスビルディング
e：竹中工務店設計部/有角昇純，久林正晴
f：竹中工務店
h：建物が密集し，緑の乏しいこの地域で，水と緑による地域のオアシス的環境づくりがテーマであった。
i：花こう岩，ステンレス，コンクリート打っ放し，クスノキ，モミジ，カツラ，アイビー
j：竹中工務店
●
a：MITSUI SOKO HAKOZAKI BUILDING
b：Chuo-ku, Tokyo
c：Mitsui Soko Office Building Corporation
e：Takenaka Corporation, Building Design Dep./Noriyoshi Arikado, Masaharu Hisabayashi
f：Takenaka Corporation
h：Being in a area where the buildings stand close together and wichi is lacking in greenery, the theme of The Creation of an Oasis Enviroment by Water and Greenery was chosen for this construction.
i：Granite, stainless steel, exposed concrete, camphor trees, Japanese maples, Katsura trees, Ivy
j：Takenaka Corporation

a：テクノウェイブ100新築工事
b：神奈川県横浜市
c：竹中工務店，佐藤工業，インテック，住友商事，
住友生命，千代田化工建設
e：竹中工務店，佐藤工業，千代田化工建設
f：竹中工務店，佐藤工業，千代田化工建設
g：ヒロフォトビルディング
h：横浜市の臨海部整備基本構想に基づく先端技術
産業誘致の事業コンペ。
i：ガラスカーテンウォール，アルミアルマイトパネル，
ラスタータイル
j：竹中工務店

●
a：TECHNOWAVE 100
b：Yokohama-shi, Kanagawa
c：Intec Inc., Sumitomo Corporation,
Sumitomo Life Insurance Company, Chiyoda
Corporation, Takenaka Corporation, Sato
Kogyo Co.,Ltd.
e：Takenaka Corporation, Sato Kogyo Co.,Ltd.,
Chiyoda Corporation
f：Takenaka Corporation, Sato Kogyo Co.,Ltd.,
Chiyoda Corporation
g：Hiro Photo Building
h：This is competition organized by Yokohama
city projected to invite leading technical
industries based on the basic idea of re-
tending to the area.
i：Glass curtain wall, aluminium almight
panels, luster tiles
j：Takenaka Corporation

a：三菱倉庫越前堀再開発
b：東京都中央区
c：三菱倉庫
e：竹中工務店設計部/矢口進，鈴木恵宜，及川義邦
f：竹中工務店
g：東京グラフィック
h：緩傾斜堤防の整備が進む隅田川のウォーターフロントの再開発親水空間と一体となった公開空地。
i：基段部：花こう岩バーナー仕上げ/高層部：乾式成型大判タイル貼り
j：竹中工務店
●
a : MITSUBISHI ECHIZENBORI DEVELOPMENT
b : Chuo-ku, Tokyo
c : Mitsubishi Warehouse & Transportaion Co., Ltd.
e : Takenaka Corporation, Building Design Dep./Susumu Yaguchi, Keigi Suzuki, Yoshikuni Oikawa
f : Takenaka Corporation
g : Tokyo Graphic
h : This public open space has been created to integrate with the main body of woter in the Sumida Waterfront Redevelopment area, of which the gently sloping banks are still under construction.
i : Ground granite steps with a banner finish, formative dry processed large tiling on the rrulti areas
j : Takenaka Corporation

a：西伊豆町総合ギャラリーパーク計画 井田子水門
展望デッキ
b：静岡県加茂郡
c：西伊豆町
d：西澤健，朝倉則幸，南和正
e：白浜力，田村賢治，佐々木賢範
f：日本軽金属
g：安川千秋，斉藤さだむ
h：田子漁港の中心に位置することから，シンボリック
で親しまれる施設をめざし展望デッキとしてデザインした。
i：機械室／アルミパネル，塔他／ステンレス，デッキ／
燻蒸板
j：GK 設計

● a : TOTAL ENVIRONMENT DESIGN PROJECT
OF NISHI-IZU-CHO, ITAGO OBSERVATION
DECK
b : Kamo-gun, Shizuoka
c : Nishi-Izu-cho.
d : Takeshi Nishizawa, Noriyuki Asakura,
Masakazu Minami
e : Tsutomu Shirahama, Kenji Tamura, Takanori Sasaki
f : Nippon Light Metal Co.,Ltd.
g : Chiaki Yasukawa, Sadamu Saito
h : Located in the center of Tago harbor, this
look-out deck has been designed with the
intention of creating a symbolic and friendly
feel to the facility.
i : Machinery room/aluminium paneling,
towers and others/stainless, deck/fumigation board
j : GK Sekkei Inc.

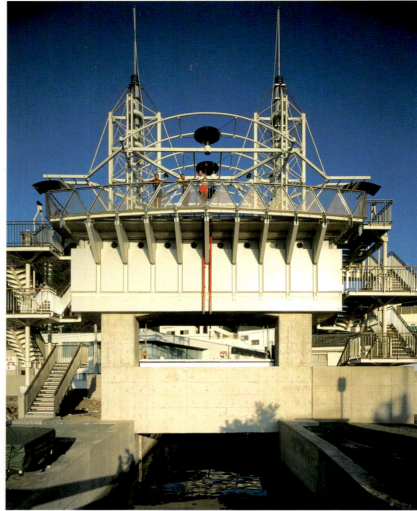

Chapter 2

レジャー環境・大型複合商業施設開発計画
Commercial and Leisure Environmental Design

「光」と「水」と「緑」がマイルドに融合したリフレッシュ・ゾーン

リーガロイヤルホテル新居浜，リーガアクアガーデン新居浜

日建設計

北四国のほぼ中央にある愛媛県第二の都市，新居浜市の住友グループ社宅跡地に，地域の振興と活性化に寄与するプロジェクトの核として，「光」，「水」，「緑」をテーマとした，アーバンリゾート指向のホテル，アクアガーデン，スポーツ施設を計画した。

建設地は市役所など公共施設や商業ビジネス地区に近く，またゴルフ場などへのレジャー基地としても便利な位置にあり，ホテル上階からは瀬戸内海や四国山脈が臨まれる。

敷地内を南北に幅6mの構内道路を通し，その両側にホテル棟とアクアガーデン棟を配置し，緑地やイベント広場などを設けて周辺道路からは十分な距離を確保し，環境を整備している。

ホテルは，シティホテルとしての機能をもち，北側市道に沿って宴会，婚礼施設のある低層棟を，後方に客室の高層棟を配置している。

アクアガーデンは，さまざまな機能のプール，老人から子供まで楽しめる各種のスパ（入浴施設），宴会場，レストラン，店舗のあるアミューズメント（遊食空間）から構成されている。これはホテルのフィットネス，ウェルネス部門を補完する機能ももち，2階に渡り廊下を設け，ホテル宿泊客がアクアガーデンへ直接アプローチできる。

また，スポーツ施設として，屋外に全天候型のテニスコート4面と，クラブハウスを設けている。

Harmonizing refresh zone with SUNSHINE, WATER and GREENERY
RIHGA Royal Hotel, Niihama & RIHGA Aqua Garden, Niihama
Nikken Sekkei Ltd

The project to build an urban resort-style hotel, aqua garden and sports facility on the former estate of the Sumitomo Group company housing area, located in Niihama city, the second largest city in Ehime prefecture which is in the center of north Shikoku, with SHINE, WATER and GREENERY as the main themes was pushed forward in order to contribute to the regional construction of a new town and to offer it added activation.

The project site is a good place for an urban resort-style hotel, since it is closely located by such facilities as the city hall, the business area and is conveniently located close to a golf course and other recreational spots, which enables the hotel to serve as a base of recreationl activities. Additionally, a view of the Seto inland sea and the Shikoku mountain range is available from the upper floors of the hotel.

A six-meter wide road was built to run through the site from north to south, and the hotel and aqua garden were arranged along it. A green space and an event square have been created in order to keep the constructions well away from the surrounding roads and therefore maintain the environment.

Serving as a city hotel, the hotel consists of a low-rise building along the city road in the north which contains the facilities necessary for banquets and wedding receptions, and the high-rise building located behind which contains the guest rooms.

The aqua garden consists of various amenities such as swimming pools having various functions, various kinds of spa facilities which can be enjoyed by a wide range of people from the very young to the very old, banquet hals, restaurants and an amusement area containing shops (a space to enjoy recreation and meals). These serve also as the fitness and wellness facilities of the hotel. The connecting corridor was built so that the guests can directly approach the aqua garden.

●配置図

ホテルの低層部は、「光」、「水」、「緑」をテーマにしたメインラウンジを中心に料飲施設及び宴会、婚礼施設を設けた。和食堂、中華レストラン、コーヒーショップ、ティーラウンジについては、どの店舗も外部に面するよう配置しリゾート性を強調した。宴会場については、四国一の規模を誇る大宴会場（820平方メートル）を中心に、中小宴会場及び婚礼施設を付帯し、地元需要に対応できるよう計画した。また、3階にはメンバーズサロン「倶楽部新居浜」を設け、エグゼクティブのためのくつろぎのスペースとした。

高層階の最上部には、新居浜周辺の眺望を生かし、フレンチレストラン及びスカイラウンジを設け、宿泊客のみならず地元客にも好評を得ている。客室は94室と少数であるが、各タイプとも「ゆとり」をテーマに高い天井、大きなベッド、ひとまわり大きいバスユニットなどのアメニティーを追求し、アーバンリゾートホテルとしての快適なホテルを演出した。

In the low-rise building, facilities for drinking, banqueting and weddings were created around the main lounge designed with SHINE, WATER and GREENERY as the main themes. The Japanese restaurant, Chinese restaurant, coffee shop and tea lounge were purposely arranged to face outwards in order to emphasize the function of a resort hotel. The hotel boasts various sizes of banquet rooms with the largest being 820 square meters; perfect to meet the local demands for wedding ceremonies. A mens' saloon going under the name of CLUB NIIHAMA has been created on the third floor as an executive space for relaxation.

Taking advantage of the wonderful view afforded of Niihama from the top floor of the high-rise building, a French restaurant and a sky lounge were opened, and both have received high acclaim not only from the hotel guests, but also from the local people.

Although the hotel only has 94 rooms, each has been designed with plenty of space and high ceilings to create the feeling of relaxation. Large beds and a size larger bath rooms also go to enforce this. The main aim was to construct a pleasant life-style in an urban hotel.

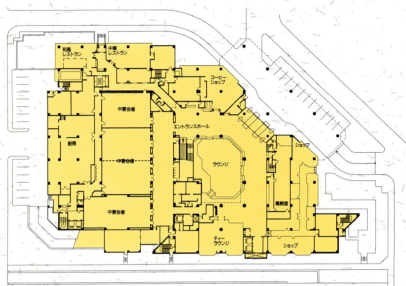

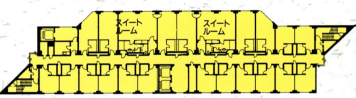

●ホテル7階平面図

プール / Swimming Pools

アクアガーデンの中心施設となるプールを計画するにあたり、「プールの遊園地化」を試みた。プールは屋外と屋内に設け、屋内プールは約2500平方メートルの床面積を有し、4分の1円形の平面形状と、最大20mもある天井高のために特異な大空間を形成している。

プールはさまざまな形状・機能をもつ「面白プール」を6種類と、ウォータースライダーなどの付属施設を設け、プールを遊園地感覚で楽しめるようにした。視覚的な演出として、プールの底面に「宇宙」をテーマとしたデザインをモザイクタイルで施した。これによりプールが巨大なオブジェとなり、屋内プールの大空間にリズムを与え、アーバンリゾートとしての華やかさ、楽しさを演出している。白一色のプールとは違い、水中での視覚(デザイン)の変化も楽しめる。また、屋内プールに熱帯性植物の植栽を行いトロピカルな空間とした。

When it came to planning the swimming pools, with hopes that it would be the main attraction of the aqua garden, it was decided to adopt an amusement-style facility. There is both an indoor and outdoor pool, and the indoor pool forms a unique space with a floor area measuring 2500 square meters.
It is a plain construction shaped in a quarter circle, and the ceiling is raised up to 20 meters at its highest point.
There are six differently shaped and functional swimming pools in total, and each has an attachement, such as a water slide, to maintain the image of amusement. As a visual production, the beds of the pools have been tiled with mosaic pictures based on the theme of the cosmos. This has turned the pools into gigantic objects and added rhythm to the space, which, in turn, has added the aura of fun and gaiety of this urban resort. A visual variation is also available in the pools which would have been missing had they been painted all in white. Tropical plants have been arranged around the indoor pool to create a tropical area.

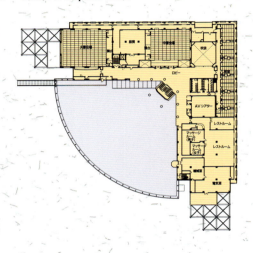

アーバンリゾートとして不可欠なスポーツ施設の一次計画として，テニスコート4面とクラブハウスを設けた。青空の下緑に囲まれた環境のなか，気持ちよく汗を流せることをコンセプトに計画を進めた。
テニスコートは全天候型でありながら土に近い自然な感覚のオムニコートとし，またテニスコートの中央部にパーゴラを設け，全体にうるおいを与えた。
テニスコート，クラブハウス，パーゴラのすべてをアースカラーとし，テニスコート全体を自然な色彩環境となるように演出した。

The priority for the construction of the sports facility, an indispensible part of an urban resort, was given to the creation of four tennis courts and a club house based on the concept that people enjoy sweating under a blue sky in an area surrounded by greenery. The court surfaces have been finished in Omni-court, an all-weather surface the color of which is close to natural earth. A pargola was placed in the center of the courts to add a touch of taste to the entire space.
The complete area of tennis courts has been arranged in a earthlike color including the actual court surfaces, the club house and the pargola.

作品名＝リーガロイヤルホテル新居浜，リーガアクアガーデン新居浜
所在地＝愛媛県新居浜市
クライアント＝リーガロイヤルホテル新居浜
ディレクター＝日建設計／近藤繁
デザイナー＝日建設計／田中秀男，甲勝之，二瓶学，伊藤節子
施工者＝住友建設／鹿島建設，J.V.（リーガアクアガーデン新居浜），鹿島建設／住友建設，J.V.（リーガロイヤルホテル新居浜）
撮影者＝SS大阪／稲住泰広
主な材料＝プール底面:50角モザイクタイル／プールサイド:100角磁器タイル
応募代表者＝日建設計

Project name=RIHGA ROYAL HOTEL, NII-HAMA & RIHGA AQUA GARDEN, NIIHAMA
Site=Niihama-shi, Ehime
Client=RIHGA Royal Hotel Niihama
Director=Nikken Sekkei/Shigeru Kondo
Designer=Nikken Sekkei/Hideo Tanaka, Katsuyuki Kabuto, Manabu Nihei, Setsuko Ito
Contractor=Sumitomo & Kajima Corporation, J.V.
Photographer=SS Osaka/Yasuhiro Inazumi
Principli materials=Pool bottoms:50mm square mosaic tiles/Pool side:100mm square ceramic tiles
Applicant=Nikken Sekkei Ltd

a：横浜伊勢佐木町ワシントンホテル
b：神奈川県横浜市
c：丸八殖産，タウン開発
d：坂倉建築研究所
e：竹中工務店設計部/臼井真，有角昇純，塩原敏
男，徳本幸雄
f：竹中工務店
g：SS東京
h：港町ヨコハマにふさわしい客船のイメージを表現し，
大通り公園のランドマークとして計画。
i：外壁：特殊面状磁器質タイル/バルコニー：スチ
ール溶融亜鉛メッキ塗装仕上げ
j：竹中工務店
●
a：YOKOHAMA ISEZAKI-CHO WASHINGTON
HOTEL
b：Yokohama-shi, Kanagawa
c：Maruhachi Shokusan Co.,Ltd., Town
Development Co.,Ltd.
d：Sakakura Associates Architects And
Engineers
e：Takenaka Corporation, Building Design
Dep./Makoto Usui, Noriyoshi Arikado, Toshio
Shiobara, Yukio Tokumoto
f：Takenaka Corporation
g：SS Tokyo
h：A passenger ship was the image chosen to
suit the well known port city of Yokohama, and
it was built as a landmark to the main street
park.
i：Exterior wall：Special ceramic tiles/
Balcony：Melted steel with a zinc plate coating
j：Takenaka Corporation

a：ビジネスホテル関屋
b：千葉県茂原市
c：関屋ビジネス
e：大成建設設計本部/前田紀貞
f：大成建設東京支店
g：篠澤建築写真事務所
h：各々の形態，色彩，素材の関係の詩をつくり出す構
造をヒントにした。
i：外装：50角モザイクタイル
j：大成建設設計本部

●

a：BUSINESS HOTEL SEKIYA
b：Mobara-shi, Chiba
c：Sekiya Business
e：Taisei Corporation Design and Proposal
Division, Norisada Maeda
f：Taisei Corporation Tokyo Branch
g：Shinozawa Architectural Photogroph Office
h：The relation among forms, colors and
materials respectively is designed by
considering poetry structure as a hint.
i：Exterior : Mosaic tile 50mm square
j：Taisei Corporation Design and Proposal
Division

a：ホテルスプリング幕張
b：千葉県千葉市
c：山形屋商事
e：竹中工務店/稲葉利行，石野連司
f：竹中工務店
g：ミヤガワ
h：光と水と緑をふんだんに取入れた，アーバンリゾートホテルの創造を目指した。
i：磁器ラスタータイル，アルミ，花こう岩，大理石，ビニールクロス
j：竹中工務店
●
a：HOTEL SPRINGS MAKUHARI
b：Chiba-shi, Chiba
c：Yamagataya Shoji Co.,Ltd.
e：Takenaka Corporation/Toshiyuki Inaba, Renji Ishino
f：Takenaka Corporation
g：Miyagawa
h：The aim was to build an urban resort hotel rich in light, water and greenery.
i：Ceramic luster tiles, aluminium, granite veneer, marble, vinyl cloth
j：Takenaka Corporation

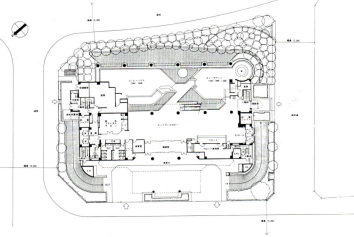

a：アムナット スクエア
b：愛知県名古屋市
c：朝日新聞社，明治生命保険，三菱商事，日本火災海上保険，竹中工務店
e：竹中工務店設計部／(建築)加藤宏生，佐藤積義，浅井康行，西崎徳房／(インテリア)安藤清，右高良樹，集山一廣
f：竹中工務店
g：村井修，SS現像所，KEN青山写真事務所
h：ホテルを核とした一街区の都心型複合施設であり，緑あふれる広場と歩道により周辺活性化を図った。
i：タイル打込Pcaコンクリート，アルミサッシ
j：竹中工務店
●

a : AMMNAT SQUARE
b : Nagoya-shi, Aichi
c : Asahi Shimbun Publishing Company, The Meiji Mutual Life Insurance Company, Mitsubishi Corporation, The Nippon Fire & Marine Insurance Co.,Ltd., Takenaka Corporation
e : Takenaka Corporation, Building Design Dep./ (Building) Hiro Kato, Sekiyoshi Sato, Yasuyuki Asai, Tokufusa Nishizaki, (Interior) Kiyoshi Ando, Yoshiki Migitaka, Kazuhiro Shuyama
f : Takenaka Corporation
g : Osamu Murai, SS Laboratory Co.,Ltd., KEN Aoyama Photo Office
h : This is a midtown-style complex facility built in a hotel block area.The creation of the square with its greenery and pathway was designed to activate the surroundings.
i : Tile cast Pca concrete, aluminium sash
j : Takenaka Corporation

a：東京ベイホテル東急
b：千葉県浦安市
c：サンプラザ
d：日本設計
e：日本設計，A.R.P.E 社（フランス・パリ）一部インテリ
ア・デザイン
f：竹中工務店，東急建設，戸田建設共同企業体
g：澤田勝良（STUDIO さわだ），岩崎和幸（カブラギスタジオ）
h：大きなガラス屋根のアトリウムの下に，南欧のリゾ
ートに見られる明るく気品のある街並みを実現。
i：床：大理石/壁：人造砂岩，大理石，鏡貼り/天
井：ガラス屋根
j：日本設計
●
a：TOKYO BAY HOTEL TOKYU
b：Urayasu-shi, Chiba
c：Sunplaza Co.,Ltd.
d：Nihon Sekkei Inc.
e：Nihon Sekkei Inc., A.R.P.E.(Paris, France)
f：Takenaka, Tokyu, Toda J.V.
g：Katsuyoshi Sawada(Studio Sawada),
Kazuyuki Iwasaki(Kaburagi Studio)
h：A pleasant and elevated place like Southern
European resorts is made under an atrium
which has a big glass roof.
i：Floor：Marble/Wall：Artificial sandstone,
marble, mirror/Ceiling：Glass roof
j：Nihon Sekkei, Inc.

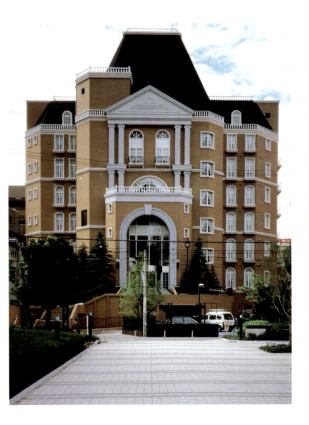

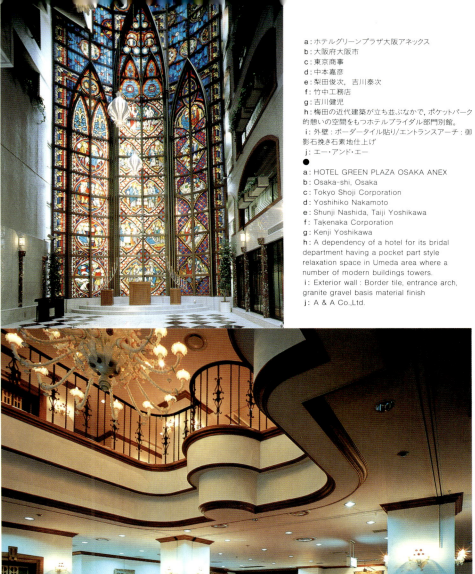

a：ホテルグリーンプラザ大阪アネックス
b：大阪府大阪市
c：東京商事
d：中本嘉彦
e：梨田俊次，吉川泰次
f：竹中工務店
g：吉川健児
h：梅田の近代建築が立ち並ぶなかで，ポケットパーク的憩いの空間をもつホテルブライダル部門別館。
i：外壁：ボーダータイル貼り／エントランスアーチ：御影石挽き石素地仕上げ
j：エー・アンド・エー
●
a：HOTEL GREEN PLAZA OSAKA ANEX
b：Osaka-shi, Osaka
c：Tokyo Shoji Corporation
d：Yoshihiko Nakamoto
e：Shunji Nashida, Taiji Yoshikawa
f：Takenaka Corporation
g：Kenji Yoshikawa
h：A dependency of a hotel for its bridal department having a pocket part style relaxation space in Umeda area where a number of modern buildings towers.
i：Exterior wall : Border tile, entrance arch, granite gravel basis material finish
j：A & A Co.,Ltd.

a：ホテル日航福岡のプランター
b：福岡県福岡市
c：九州勧業，ホテル日航福岡
d：鹿島建設インテリアデザイン部
e：イワオ・ベセラ・デザインチーム
h：国際化する福岡に建てられた都市型ホテル内のシンプルなツボ型のプランター。
i：磁器
j：岩尾磁器工業
●
a：HOTEL NIKKO FUKUOKA PLANTER
b：Fukuoka-shi, Fukuoka
c：Kyushu Kangyo Co.,Ltd., Hotel Nikko Fukuoka
d：Kajima Corporation, Interior Design Division
e：Iwao Jiki Vessela Design Team
h：A simple designed jar style planter in an urban hotel built in Fukuoka which is becoming international.
i：Vitreous
j：Iwao Jiki Kogyo Co.,Ltd.

a：関東地区デパート健康保険組合鬼怒川保養所「せせらぎ荘」
b：栃木県今市市
c：関東地区デパート健康保険組合
d：三宅健司
e：関原聡，西田正則
f：五洋建設
g：門馬金昭
h：鬼怒川沿いの自然の緑の中に建物を溶けこませ、自然公園といった周辺環境を考慮した保養施設。
i：シラカシ，ヤマモミジ，ヤマザクラ，ニシキギ，クマザサ 等
j：日建設計
●

a : KINUGAWA REST HOME, SESERAGI-SO
b : Imaichi-shi, Tochigi
c : Kanto Department Store Health Insurance Society
d : Kenji Miyake
e : Satoru Sekihara, Masanori Nishida
f : Penta Ocean
g : Kaneaki Monma
h : This health faciliity has been built to integrate itself with the green nature along the banks of the Kinugawa River as it is located near the natural park.
i : White oak, mountain maple, mountain cherry trees, winged spindle trees, striped bamboo, etc.
j : Nikken Sekkei Ltd.

a：神戸製鋼所ラグビー部クラブハウス
b：兵庫県神戸市
c：神戸製鋼所
e：藤縄正俊，船橋昭三
f：大林組
g：ナトリ光房
h：神鋼ラグビー部の日本選手権 V1の記念性をシンボライズする形を表現した。
i：せっ器質タイル，アルミサッシ，ミラーガラス
j：大林組本店一級建築士事務所
●

a：KOBE STEEL RUGBY FOOTBALL CLUB HOUSE
b：Kobe-shi, Hyogo
c：Kobe Steel Co.,Ltd.
e：Masatoshi Fujinawa, Shozo Funahashi
f：Obayashi Corporation
g：K.Natori Studio
h：It is designed to symbolize in commemoration of Kobe Steel Rugby football team's V1 for the Japanese championship.
i：stone ware, aluminum sash, mirror glass
j：Obayashi Corporation

a：サウスサイドコート
b：大阪府大阪市
c：美津濃
e：三谷幸司，林浩二，梶原一幸（大林組）
f：大林組
g：東出清彦/SS 大阪
h：来館者を主役としてひきたてるため，薄いグレーを
基調として明るく健康的な雰囲気づくりを目指した。
i：45角モザイクタイル/ステンレス/アルミ/タイルカー
ペット/ダイナミックフロア
j：大林組本店建築設計第1部
●

a：SOUTH SIDE COURT
b：Osaka-shi, Osaka
c：Mizuno Corporation
e：Koji Mitani, Koji Hayashi, Kazuyuki Kajihara
(Obayashi Corp.)
f：Obayashi Corporation
g：Kiyohiko Higashide/SS Osaka
h：It is designed to have a bright and healthy
atmosphere with a coloring based on light
gray in order to set off visitors as they are
stars.
i：Mosaic tile 45 mm square, stainless steel,
aluminum, tile carpet, dynamic floor
j：Obayashi Corporation, Design Departments,
No.1

a：ニドム リゾート
b：北海道苫小牧市
c：ザ・ニドム
d：佐々木桂
e：塚本武司，佐藤正徳，金山正志，木田輝夫
f：乃村工藝社
g：エイトプロダクション
h：豊かな自然の中でのリゾートライフをログハウス，
ホテル施設，スポーツ施設を通し提案している。
i：石/シベック，カピスト，御影石，インド砂岩，チー
ク，マジックコート，シルバーパイン
j：乃村工藝社
●

a : NIDOM RESORT
b : Tomakomai-shi, Hokkaido
c : The Nidom Corporation
d : Katsura Sasaki
e : Takeshi Tsukamoto, Masanori Sato, Masashi
Kanayama, Teruo Kida
f : Nomura Co.,Ltd.
g : Eight Production
h : Resort life amongst the wealth of nature
has been proposed through log cabins, hotels
sports facilities.
i : Stone/sveck capist, granite stone, indian
sagrock, teak, magic coating, shilled pine
j : Nomura Co.,Ltd.

a：軽井沢リゾートヴィラ4号館
b：群馬県吾妻郡
c：東京商事
d：中本嘉彦
e：吉川泰次, 吉本浩
f：浅沼組東京本店
h：雄大な好環境リゾートゾーンの中に建つ英国伝統の木組家屋風チューダー様式のホテル。
i：外壁：模様吹き付け塗料（ジョリパット）吹き付け/ロビー・外部床：レンガタイルパターン貼り
j：エー・アンド・エー
●

a : KARUIZAWA RESORT VILLA # 4
b : Agatsuma-gun, Gunma
c : Tokyo syouji Corporation
d : Yoshihiko Nakamoto
e : Taiji Yoshikawa, Hiroshi Yoshimoto
f : Asanumagumi Corporation
h : A wood construction housing type tuder style hotel which built in a grand nice environmental resort zone.
i : Exterior wall : Pattern spray paint.Lobby & outside floor : Brick tile
j : A & A Co.,Ltd.

a：藍澤証券白樺研修センター
b：長野県北佐久郡
c：藍澤証券
e：大成建設設計本部/小笠原祥仲
f：大成建設北信越支店
g：藤江猛（フォトセンター）
h：国定公園内の美しい自然に溶け込み、ヒューマンスケールを維持した外・内観を求めた。
i：磁器質タイル、フッ素樹脂焼付アルミ屋根、黒御影石
j：大成建設設計本部
●

a：SHIRAKABA TORAINING CENTER OF AIZAWA SECURITIES Co.,Ltd.
b：Kita-Saku-gun, Nagano
c：Aizawa Securities Co., Ltd
e：Taisei Corporation Design and Proposal Division, Yoshinaka Ogasawara
f：Taisei Corporation Kita-Shinetsu Branch
g：Takeshi Fujie (Phote Center)
h：Exterior and interior of a building is designed to keep touches of humanity with integration the building into the beautiful nature inside of a national park.
i：Vitreous tile, aluminum roof baked fluororesin, black granite
j：Taisei Corporation Design and Proposal Division

a：ユーラシア404
b：北海道小樽市
c：松井力
e：大成建設札幌支店設計部/竹井俊介
f：大成建設札幌支店
g：安達治
h：RC 打っ放しにした建物と石敷きの庭園は海と周囲の風景に同化し、プライマリーな空間を提供する。
i：コンクリート打っ放し、アルミパンチング、スチール、強化ガラス、フローリング、じゅうたん
j：大成建設設計本部
●

a：EURASIA 404
b：Otaru-shi, Hokkaido
c：Tsutomu Matui
e：Taisei Corporation Sapporo Branch, Shunsuke Takei
f：Taisei Corporation Sapporo Branch
g：Osamu Adachi
h：A building with RC exposed and a garden with stone pavement offer a primary space by assimilating with the sea and surrounding scenery.
i：Exposed concrete, aluminum punching, steel, safety glass, flooring, carpet
j：Taisei Corporation Design and Proposal Division

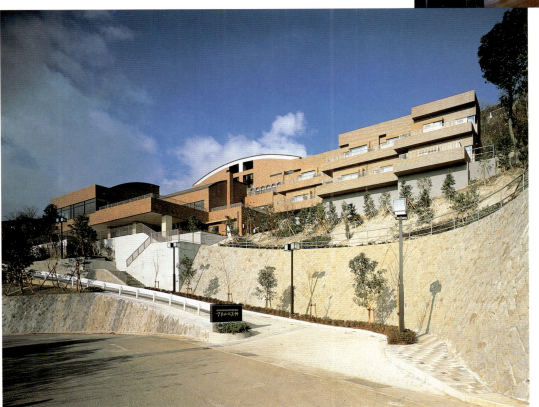

a：日本経済新聞社健康保険組合，伊豆山保養所
b：静岡県熱海市
c：日本経済新聞社健康保険組合
e：大成建設本部/田平徹，桜井篤信
f：大成建設横浜支店
g：三輪写真
h：斜面に建つ建物，周囲の山並みにあった曲線を持つボールト屋根，男性的な外壁に対し女性的な内装。
i：銅板，特殊レンガタイル
j：大成建設設計本部
●

a：NIHON KEIZAI SHINBUNSHA IZUSAN RESORT HOUSE
b：Atami-shi, Sizuoka
c：Nihon Keizai Shinbunsha
e：Taisei Corporation Design and Proposal Division, Toru Tahira, Atsunobu Sakurai
f：Taisei Corporation Yokohama Branch
g：Miwa Photo Studio
h：A building standing on a sloping surface.A vault having a curved line fitting surrounding mountains.Feminine interior as against masculine exterior wall.
i：Steel plate, special brick tile
j：Taisei Corporation Design and Proposal Division

a：白馬コルチナリゾート・ヴィラ
b：長野県北安曇郡
c：日本オーナーズクラブ
d：中本嘉彦
e：梨田俊次，吉本浩
f：大成建設
g：佐竹勝也
h：広大なスキーリゾートのなかの若者向けホテル。英国伝統のチューダー様式でまとめた巨大施設。
i：外壁：模様吹き付け塗材（ジョリパット）吹き付け／ロビー・外部床：レンガタイルパターン貼り
j：エー・アンド・エー
●

a：HAKUBA CORTINA RESORT VILLA
b：Kita-Azumi-gun, Nagano
c：Nihon Owners Club Corporation
d：Yoshihiko Nakamoto
e：Shunji Nashida, Hiroshi Yoshimoto
f：Taisei Kensetu Corporation
g：Katsuya Satake
h：A hotel aiming at young people in a huge ski resort.It is a massive facility coordinated to be tudor style of English traditional.
i：Exterior wall : Pattern spray painting/Lobby & outside floor : Brick tile
j：A & A Co.,Ltd.

a：国際交流会館　芝蘭会館
b：京都府京都市
c：芝蘭会
e：竹中工務店設計部/野村充，薄田学
f：竹中工務店
g：村井修
h：内外空間の重層性という京の町家がもつ伝統的空
間形式の現代へのトランスフォーメーション。
i：磁器タイル，花こう岩，コンクリート打っ放し，アル
ミメッキ鋼板，大理石
j：竹中工務店
●

a：SHIRAN KAIKAN
b：Kyoto-shi, Kyoto
c：Shirankai
e：Takenaka Corporation, Building Design
Dep./Mitsuru Momura, Manabu Susukida
f：Takenaka Corporation
g：Osamu Murai
h：A modern transformation of the traditional
style of tradesman's house in Kyoto of which
thez space within and without is stratified.
i：Ceramic tiles, granite veneer, exposed
concrete, aluminium plated steel boaeds,
marble
j：Takenaka Corporation

a：横浜アリーナ
b：神奈川県横浜市
c：横浜アリーナ
e：竹中工務店設計部/榎本和夫，二宮孝，森川昇，
茶谷明男
f：竹中工務店
g：渡辺洋美
h：市政百周年事業の一環として，フレキシビリティに
富む空間・機能を有するホールとして計画。
i：外壁・磁器質二丁掛タイル，アルミカーテンウォー
ル，熱線反射ガラス
j：竹中工務店
●

a : YOKOHAMA ARENA
b : Yokohama-shi, Kanagawa
c : Yokohama Arena Co.,Ltd.
e : Takenaka Corporation, Building Design
Dep./Kazuo Kazuo, Takashi Ninomiya, Noboru
Morikawa, Akio Chatani
f : Takenaka Corporation
g : Hiromi Watanabe
h : As part of the city's 100 year anniversary
project, this hall was constructed with a serene
flexiblity both space-wise and function-wise.
i : Exterior wall : Ceramic tilles, aluminium
curtain wall, heat-reflecting glass
j : Takenaka Corporation

a：横浜そごう雲見迎賓館
b：静岡県賀茂郡
c：七博
e：大成建設設計本部/西牟田章夫
f：大成建設，土屋建設共同企業体
g：三輪哲士
h：外人客を想定し，デザイン基調を内外装とも和洋折衷型で整えた。緑青銅板を効果的に使用。
i：二丁掛磁器質タイル，銅板
j：大成建設設計本部
●
a：YAKOHAMA SOGO KUMOMI GUEST HOUSE
b：Kamo-gun, Shizuoka
c：Shichihiro
e：Taisei Corporation Design and Proposal Division, Akio Nishimuta
f：Taisei, Tsuchiya J.V.
g：Tetsuji Miwa
h：Supporting foreign guests, a tone of design is coordinated of internal and external to be a blending of Japanese and Western style. Aquamarine copper plates are used effectively.
i：Double sized vitreous tiles, copper plate
j：Taisei Corporation Design and Proposal Division

a：愛宕原ゴルフ倶楽部 クラブハウス
b：兵庫県宝塚市
c：播磨興産
e：竹中工務店設計部/坂根恭司，白木孝志
f：竹中工務店
g：古田雅文
h：自然石（庵治石）とコンクリート打っ放しの質量感ある素材を外装にもちい，環境との調和を図った。
i：床：タイルカーペット/壁：クロスペンキ，一部庵治石/天井：岩綿吸音板ペンキ
j：竹中工務店
●

a：ATAGOHARA GOLF CLUB, CLUB HOUSE
b：Takarazuka-shi, Hyogo
c：Harima Kosan Co.,Ltd.
e：Takenaka Corporation, Building Design Dep./Kyoji Sakane, Takashi Shiraki
f：Takenaka Corporation
g：Masafumi Koda
h：Natural stone and voluminous amounts of exposed concrete have been used as an exterior decoration in order to harmonize with the surroundings.
i：Foor : Tile carpeting/wall : Cloth paint, a little Anji stone/Ceiling : acoustic asbestos board paint
j：Takenaka Corporation

a：ビッグワンカントリークラブ信楽コース，クラブハウス
b：滋賀県信楽町
c：大陽興業
e：大成建設大阪支店設計部/美濃吉昭，浦山哲
f：大成建設
g：SS大阪
h：和風感覚をテーマとしたクラブハウス。一文字本瓦と白壁の外観，モダンなインテリアが特徴。
i：一文字日本瓦，米松練付合板，花こう岩
j：大成建設設計本部
●
a：BIG ONE COUNTRY CLUB SHIGARAKI, CLUB HOUSE
b：Shigaraki-cho, Shiga
c：Taiyo Kogyo Co.,Ltd.
e：Taisei Corporation Osaka Branch Design Division, Yoshiaki Mino, Satoshi Urayama
f：Taisei Corporation
g：SS Osaka
h：A club house of which theme is Japanesque.The features are a modern interior decoration and exterior consisting of straight line Japanese tiles and white walls.
i：Straight line Japanese tiles, douglas spruce veneering plywood, granite
j：Taisei Corporation Design and Proposal Division

a：東京よみうりカントリークラブ　クラブハウス
b：東京都稲城市
c：よみうりランド
e：大成建設設計本部/永井英敏，安部充
f：大成建設
g：高井潔
h：重なりあう緑のヴォールト屋根に包まれた，明るく個性的なクラブハウス。
i：緑青銅板，二丁掛磁器質タイル，アルミパネル，成型岩綿吸音板，セラキューブ，じゅうたん
j：大成建設設計本部
●
a：TOKYO YOMIURI COUNTORY CLUB CLUB HOUSE
b：Inagi-shi, Tokyo
c：Yomiuri Land Co.,Ltd.
e：Taisei Corporation Design and Proposal Division, Hidetoshi Nagai, Mitsuru Abe
f：Tausei Corporation
g：Kiyoshi Takai
h：A bright and unique club house surrounded by green vaulted roof which are overlapped each other.
i：Aquamarine copper sheet, double sized vitreous tile, aluminum panel, shaped rock wool acoustic tile, ceramic-cube, carpet
j：Taisei Corporation Design and Proposal Division

a：京葉国際カントリークラブ，クラブハウス
b：千葉県千葉市
c：京葉国際カントリークラブ
e：大成建設設計本部/保科朝男，春田義行
f：大成建設東京支店
g：三輪晃士
h：3棟が織りなす銅板葺の寄棟屋根と，土色割肌タイ
ルの外壁で30年を経た環境に調和させた。
i：銅板，割肌二丁掛タイル，インド砂岩，ナラ練付，
カーペット
j：大成建設設計本部
●
a：KEIYO KOKUSAI COUNTRY CLUB, CLUB
HOUSE
b：Chiba-shi, Chiba
c：Keiyo Kokusai Country Club
e：Taisei Corporation Design and Proposal
Division, Asao Hoshina, Yoshiyuki Haruta
f：Taisei Corporation Tokyo branch
g：Koshi Miwa
h：Three buildings of which roof are sheet
copper hip roof and the exterior wall made of
earth colored chopped face tiles is
harmonized with surroundings in 30 years' time.
i：Copper sheet, double sized chopped face
tiles, Indian sandstone, oak veneering, carpet
j：Taisei Corporation Design and Proposal Division

a：花の木ゴルフクラブ，クラブハウス
b：岐阜県瑞浪市
c：名古屋テレビ開発
e：大成建設名古屋支店設計部/伊井伸
f：大成建設
g：センター・ホト
h：方形屋根4棟を45度振りながら連らねて水平線を
強調，30mの塔がのびバランスを保つ。
i：花こう岩，スタッコ，屋根カラーアルミ
j：大成建設設計本部
●
a : HANANOKI GOLF CLUB, CLUB HOUSE
b : Mizunami-shi, Gifu
c : Nagoya TV Development Ltd.
e : Taisei Corporation Nagoya Branch Design
Division, Shin Ii
f : Taisei Corporation
g : Senter Hoto
h : Four buildings having square hipped roof
are sprinkled with and ranged to emphasize
the horizon.30-meter-tower extend and keep
its balance.
i : Granite, stucco, roof color aluminum
j : Taisei Corporation Design and Proposal
Division

a：東広野ゴルフ倶楽部クラブハウス
b：兵庫県三木市
c：三津田開発
d：石阪春生
e：大成建設大阪支店設計部/美濃吉昭，富田博
f：大成建設大阪支店
g：伸和実業
h：自然と素材のラフなテクスチャーと，素朴な形態の構成に依る，風格のある空間創り。
i：屋根：天然スレート/外壁：花こう岩割石積，磁器タイル
j：大成建設設計本部
●

a : HIGASHI-HIRONO GOLF CLUB, CLUB HOUSE
b : Miki-shi, Hyogo
c : Mitsuda Kaihatsu Corporation
d : Haruo Ishizaka
e : Taisei Corporation Osaka Brabch Design Division, Yoshiaki Mino, Hiroshi Tomita
f : Taisei Corporation Osaka Branch
g : Shinwa Jitsugyo Corporation
h : Distinctive space is made by composing rough texture and artless form of natural materials.
i : Roof : State roofing.Exterior wall : Granite rag-stone masonry, vitreous tile
j : Taisei Corporation Design and Proposal Division

a：東広野ゴルフ倶楽部クラブハウス
b：兵庫県三木市
c：三津田開発
d：石阪春生
e：大成建設大阪支店設計部/美濃吉昭，富田博
f：大成建設大阪支店
g：伸和実業
h：自然と素材のラフなテクスチャーと，素朴な形態の構成に依る，風格のある空間創り。
i：屋根：天然スレート/外壁：花こう岩割石積，磁器
タイル

a：蓼科ブライトン倶楽部
b：長野県茅野市
c：長谷工コーポレーション
e：梅原二六（空間計画研究所），歌田次男（長谷工コ
ーポレーションエンジニアリング事業部）
f：諏訪土木建設
g：ヤジマカメラ
h：別荘のゆとりとホテルのサービスをミックスさせた
高級会員制リゾート倶楽部。
i：コンクリート打っ放し、チリ仕上げ，焼すぎレンガタイ
ル，ジョリパットローラー，銅板
j：長谷工コーポレーションエンジニアリング事業部
●
a：TATESHINA BRIGHTON CLUB
b：Chino-shi, Nagano
c：Haseko Corporation
e：Niroku Umehara（Institute of Space
Planning），Tsugio Utada（Haseko Corporation,
Architectutr & Engneering Dept.）
f：Suwa Doboku Kenchiku Co.,Ltd.
g：Yajima Kamera
h：A high-grade membership resort club
mixing ease feeling of a villa and hotel-like service.
i：Exposed concrete, clinker brick tile, copper plate
j：Haseko Corporation Engineering Dept.

a：ショッパーズプラザ新浦安
b：千葉県浦安市
c：新商，浦安中央開発
d：長谷工コーポレーション，アール・アイ・エー
e：企画：ダイエーリアルエステート&長谷工コーポレーション，監理：長谷工・フジタ建設共同企業体設計監理室，内装：ダイエーリアルエステート浦安内装監理室，設計協力：乃村工芸社
f：長谷工・フジタ建設 J.V.
h：ダイエーを核店舗に240の専門店，スポーツ施設，公共施設，オフィス，駐車場を備えた大規模複合施設。
i：フッ素樹脂塗装大型ALC板，大理石，タイルカーペット，岩綿吸音板
j：長谷工コーポレーションエンジニアリング事業部
●
a：SHOPPERS PLAZA SHIN-URAYASU
b：Urayasu-shi, Chiba
c：Shinsho Co.,Ltd.Urayasu Chuoh Kaihatsu Co., Ltd
d：Haseko Corporation, R.I.A.
e：Plan：DRE and Haseko Corporation/ Control：Haseko Corporation, Fujita, J.V. Planning Control Section/Interier Coordinator Section DRE Construction Dept.Urayasu/ Assistanat Coordinator：Nomura Kogeisha
f：Haseko Corporation, Fujita J.V.
h：A large scale multiple facilities establishing 240 specialty stores with Daiei as the center, a gymnasium, public facilities, offices and a motor pool.
i：Fluorine resin coating large-scale ALC panels, marble, tile carpet, rock wool sound absorbent
j：Haseko Corporation Engineering Dept.

a：ショッパーズプラザ新浦安
b：千葉県浦安市
c：ダイエーリアルエステート
d：乃村工藝社／植草孝，芝田良治，佐藤孝幸
e：乃村工藝社／横塚潔，竹尾健一，荘司訓由，川村清志，西山千明，倉光瑞枝，椎名清作，石坂浩一，切道敏郎，建入通，木田和秀，武石正宣
f：乃村工藝社
g：ジャプロ
h：東京駅から15分，快適な住居と商業，交通そして業務のバランスのとれた街づくりを目指す浦安 AMC 地区の商業集積の核施設として計画・推進された。
j：乃村工藝社

●
a：SHOPPERS PLAZA SHINURAYASU
b：Urayasu-shi, Chiba
c：Daiei Real Estate Co.,Ltd.
d：Nomura Co.,Ltd./Takashi Uekusa, Ryoji Shibata, Takayuki Sato
e：Nomura Co.,Ltd./Kiyoshi Yokozuka, Kenichi Takeo, Kuniyoshi Shoji, Kiyoshi Kawamura, Chiaki Nishiyama, Mizue Kuramitu, Seisaku Shiina, Koichi Ishizaka, Toshiro Kirimichi, Toru Tateiri, Kazuhice Kida, Masanori Takeishi
f：Nomura Co.,Ltd.
g：Japro Co.,Ltd.
h：This construction was planned and thrust forward as a core facility of busimess integration, aiming at well balanced town planning living-wise, business-wise and commutation-wise owing to it being located only 15 minute away from Tokyo.
j：Nomura Co.,Ltd.

a：京阪京橋駅ビル
b：大阪府大阪
c：京阪電気鉄道
e：竹中工務店設計部/村上元康，潮先克，東政美
f：竹中工務店，京阪建設 J.V.
g：大島勝寛，東出清彦
h：三角形の敷地を生かしたシャープな外観とし，駅の
コンコースに面して2階分吹抜の「ガレリア」を設けた。
i：タイル打込 Pca 板，結晶化ガラス，アルミ，熱線
反射ガラス
j：竹中工務店
●
a：KEIHAN KYOBASHI STATION BUILDING
b：Osaka-shi, Osaka
c：Keihan Electric Railway Co.,Ltd.
e：Takenaka Corporation, Building Design
Dep./Motoyasu Murakami, Masaru Shiosaki,
Masami Higashi
f：Takenaka Corporation, Keihan Corporation J.V.
g：Katsuhiro Oshima, Kiyohiko Higashide
h：Marking the best use of the triangluar
shape that faces the concourse of the station,
a solid two-storied GALERIE has been created.
i：tiles cast in Pca board, crystalized glass,
aluminium, heat-reflecting glass
j：Takenaka Corporation

a：ポラール一番館
b：兵庫県神戸市
c：新星和不動産
e：藤縄正俊，船橋昭三，大橋眞由美
f：大林組
g：織田努写真事務所
h：切妻屋根を採用し建物を分節することで，周辺街並のシルエットとの調和とシンボル性を狙った。
i：吹き付けタイル，ガラスカーテンウォール
j：大林組本店一級建築士事務所
●
a：PORAL ICHIBANKAN
b：Kobe-shi, Hyogo
c：Sinseiwa Fudosan Co.,Ltd.
e：Masatoshi Fujinawa, Shozo Funahashi, Mayumi Ohashi
f：Obayashi Corporation
g：Tsutomu Oda Studio
h：A building is designed with dividing it into some parts by adopting a gable roof to symbolize and harmonize with a silhouette of surroundings of the city.
i：Sprayed tile, glass curtain wall
j：Obayashi Corporation

a：岩鋳キャスティングワークス
b：岩手県盛岡市
c：岩鋳
d：朝永徹一
e：千葉修良，テイ・グラバー
f：建築：中野組東北支店，内装：大丸装工事業部，
サイン：テイ・グラバー
g：藤戸充，安部順
h：南部鉄器の歴史と将来への展望を明らかにする文
化発信基地を目指し，企画・設計・施工した。
i：H鋼，プレート，トラス材，テンション材，ガラス材
j：テイ・グラバー
●
a：IWACHU CASTING WORKS
b：Morioka-shi, Iwate
c：IWACHU
d：Tetsuichi Tomonaga
e：Nagayoshi Chiba, T.Glover Co.,Ltd.
f：Nakano-Gumi, Daimaru, T.Glover Co.,Ltd.
g：Mitsuru, Fujito, Jun Abe
h：It is planned, designed and carried out with
aiming to be a culture dispatching station
which make clear the history of Nanbu iron
vessels and the views for the future.
i：H-section, plate truss material, tension
material, glass material
j：T.Glover Co.,Ltd.

a：伊勢丹新宿店2階「ザ・メッセージ」
b：東京都新宿区
c：伊勢丹
d：乃村工藝社/金子太郎
e：今井彰，須藤純治，岩崎正
f：伊勢丹施設部
g：吉田幸世
h：好感度ヤング対象の売場，導入の難しい躯体条件（旧ストック）を広がり感と回遊性をコンセプトに構築。
i：テラゾー，アンティコスタッコ，ボンスエード，SUSミラー
j：乃村工藝社

●
a：ISETAN SHINJUKU 2F THE MESSAGE
b：Shinjuku-ku, Tokyo
c：Isetan Co.,Ltd.
d：Nomura Co.,Ltd./Taro Kaneko
e：Akira Imai, Junji Sudo, Tadashi Iwasaki
f：Isetan Co.,Ltd.
g：Yukio Yoshida
h：Young people and polular markets.With expansion and circular touring as the main concept, the condition of building frame that is difficult to induce has been overcome.
i：Terrazzo, anticostucco, bonds aid, SUS mirroring
j：Nomura Co.,Ltd.

a：コーナーハウス ドゥ
b：兵庫県神戸市
c：有限会社青柳 取締役 宇津隆平
e：藤縄正俊，船橋昭三
f：大林組
g：東出清彦
h：塔に代表される神戸，北野の伝統的イメージの継承と，「北野の風景に溶け込むシルエット」が狙い。
i：せっ器質タイル，アルミサッシ，熱線吸収ガラス
j：大林組本店一級建築士事務所
●
a：CORNER HOUSE DO
b：Kobe-shi, Hyogo
c：Aoyagi Ltd., Ryuhei Utsu
e：Masatoshi Fujinawa, Shozo Funahashi
f：Obayashi Corporation
g：Kiyohiko Higashide
h：It is designed to succeed to the traditional image of Kitano Kobe represented towers and to express the image of A silhouette integrating into landscape of Kitano.
i：stone ware, aluminum sash, heat absorbing glass
j：Obayashi Corporation

a：MYCAL 本牧
b：神奈川県横浜市
c：ニチイ
e：大成建設設計本部/清水一夫，伊藤崇洋，森繁文
雄，河村光男，人見修
f：大成建設横浜支店
g：三輪写真
h：横浜市とデベロッパーニチイの意向が合意した開
発建築計画と街づくりの一環。
i：外壁：PC板吹き付けタイル
j：大成建設設計本部

●

a：MYCAL HONMOKU
b：Yokohama-shi, Kanagawa
c：Nichii Corporation
e：Taisei Corporation Design and Proposal
Division, Kazuo Shimizu, Takahiro Itoh, Fumio
Morishige, Mitsuo Kawamura, Osamu Hitomi
f：Taisei Corporation, Yokahama Branchi
g：Miwa Shasin
h：It forms a part of the development building
and town plan which Yokohama city's intention
and a developer Nichii's meet.
i：Exterior wall : PC plate sprayed tile
j：Taisei Corporation Design and Proposal Division

a：MYCAL 本牧
b：神奈川県横浜市
c：ニチイ
e：大成建設設計本部/清水一夫，伊藤崇洋，森繁文
雄，河村光男，人見修
f：大成建設横浜支店
g：三輪写真
h：横浜市とデベロッパーニチイの意向が合意した開
発建築計画と街づくりの一環。
i：外壁：PC板吹き付けタイル
j：大成建設設計本部

a：ブリックハウス・あしび
b：千葉県千葉市
c：千葉そごう，あしび
e：大成建設設計本部/西田幸男
f：大成建設
g：三輪晃士
h：レンガ積みの外観と和洋混合のインテリアで構築
されたアンティックなカフェ＆レストラン。
i：英国産レンガ，御影石，ロートアイアン，米マツ，
土壁
j：大成建設設計本部
●
a：BRICK HOUSE ASHIBI
b：Chiba-shi, Chiba
c：Chiba Sogo Co.,Ltd., Ashibi Co.,Ltd.
e：Taisei Corporation Design and Proposal
Division, Yukio Nishida
f：Taisei Corporation
g：Koji Miwa
h：An antique cafeteria restaurant constructed
by brick masonry exterior and interior
decoration blending of Japanese and Western
styles.
i：Brick made in the UK,
granite, lout iron,
douglas spruce, mud wall
j：Taisei Corporation Design
and Proposal Division

a：POLA 札幌南1条ビル
b：北海道札幌市
c：忍総業
f：竹中工務店
g：新津写真，吉村行雄
h：シンプルでシンボリックな，洗練された美しさを感じ
る感性豊かな複合ビル。
j：竹中工務店
●
a：POLA SAPPORO MINAMI ICHIJO BUILDING
b：Sapporo-shi, Hokkaido
c：Shinobu Sogyo
f：Takenaka Corporation
g：Niitsu Photo, Yukio Yoshimura
h：A complex building with an impression of
simple and symbolically refined beauty.
j：Takenaka Corporation

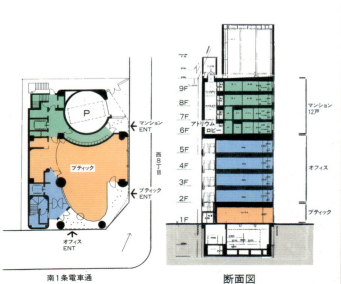

断面図

a：シエスタ伊豆高原
b：静岡県伊東市
c：長谷工コーポレーション　ウエル・センター
d：谷岡義高
e：池上一夫，永井幹子
f：東急建設
g：東京 D.P.E
h：南欧風のモダンな外観に自然素材をバランスよく組み合わせ、心地よい緊張感を演出。
i：マジックコート，チーク，赤レンガ，テラコッタタイル
j：長谷工コーポレーションエンジニアリング事業部
●
a：SIESTA IZU KOGEN
b：Ito-shi, Shizuoka
c：Haseko Corporation Well-Center
d：Yoshitaka Tanioka
e：Kazuo Ikegami, Mikiko Nagai
f：Tokyu
g：Tokyo D.P.E.
h：Southern European style.It is designed to express a comfortable tense atmosphere combining natural materials in well-balanced.
i：Magic coat, teak, red brick, terra cotta tile
j：Haseko Corporation Engineering Dept.

a：IMP インターナショナルマーケットプレイス
b：大阪府大阪市
c：松下アイ・エム・ピー
d：郷力憲治
e：船津茂，岡田哲哉，中村裕輔，福山秀親，三村拓也，保川彰宏
f：MID ビルサービス
g：村瀬武男
h：「パスポートのいらない外国」IMP は，独自の流通システムによる海外直輸入ショッピングモール。
i：大理石本磨模様貼り一部大版タイル貼り，LED パネルライト埋込テラゾ貼り
j：乃村工藝社

●

a：THE INTERNATIONAL MARKET PLACE
b：Osaka-shi, Osaka
c：Matsushita IMP Co.,Ltd.
d：Kenji Goriki
e：Shigeru Funatsu, Tetsuya Okada, Yusuke Nakamura, Hidechika Fukuyama, Takuya Mimura, Akihiro Yasukawa
f：MID Building Service Co.,Ltd.
g：Takeo Murase
h：IMP called A foreign land where you can pass through without a passport is a direct import shopping mall by individual distribution system.
i：Polished marble/Terrazzo recessed LED panel light
j：Nomura Co.,Ltd.

a：調布駅北口ビル（調布パルコ）
b：東京都調布市
c：パルコ
d：増田通二
e：大成建設設計本部/吉田進，金納弘一
f：大成建設
g：井上光伸
h：「郊外性」と「都会性」の融合をめざした複合型商業
施設（物販＋ホテル＋映画館）。
i：低層部：モザイクタイル＋コンクリート打っ放し，高
層部：コンクリートPC板＋アルミカーテンウォール
j：大成建設設計本部
●
a：CHOFU PARCO
b：Chofu-shi, Tokyo
c：Parco Co.,Ltd.
d：Tsuji Masuda
e：Taisei Corpration Design and Proposal
Division, Susumu Yoshida, Koichi Kanno
f：Taisei Corporation
g：Mitsunobu Inoue
h：A multiple commercial establishment aiming
to harmonize SUBURBIA and URBANITY
(ships, a hotel and movie theater.)
i：Lower part：Mosaic tiles, exposed
concrete/Higher part：Concrete PC panels,
aluminum curtain wall
j：Taisei Corporation Design and Proposal Division

a：グランディ新横浜
b：神奈川県横浜市
c：グランディ
d：友水孝一
e：鍵山一生，中村秀男
f：大成建設，丹青社
g：カブラギ写真事務所
h：12店舗の総合飲食店ビル。音楽と食文化の融合を
コンセプトに，ハープの音色の美しさをもくろむ。
i：外壁＝2種類の花こう岩/特注アルミカーテンウォー
ル/アルミ，ステンレスパネル
j：像建築設計事務所
●
a：GRANDY SHINYOKOHAMA
b：Yokohama-shi, Kanagawa
c：Grandy Co.,Ltd.
d：Koichi Tomomizu
e：Kazuo Kagiyama, Hideo Nakamura
f：Taisei Corporation, Tanseisha Co.,Ltd.
g：Kaburagi Photo Office
h：A complex food business building
consisting of 12 restaurants.It is designed with
the concept of harmonizing music and food
culture to express the beautiful tone of a harp.
i：Exterior wall : Two kind of granite, custom
aluminum curtain, aluminum and stainless steel
panel
j：Zou, Architects & Associates, Inc.

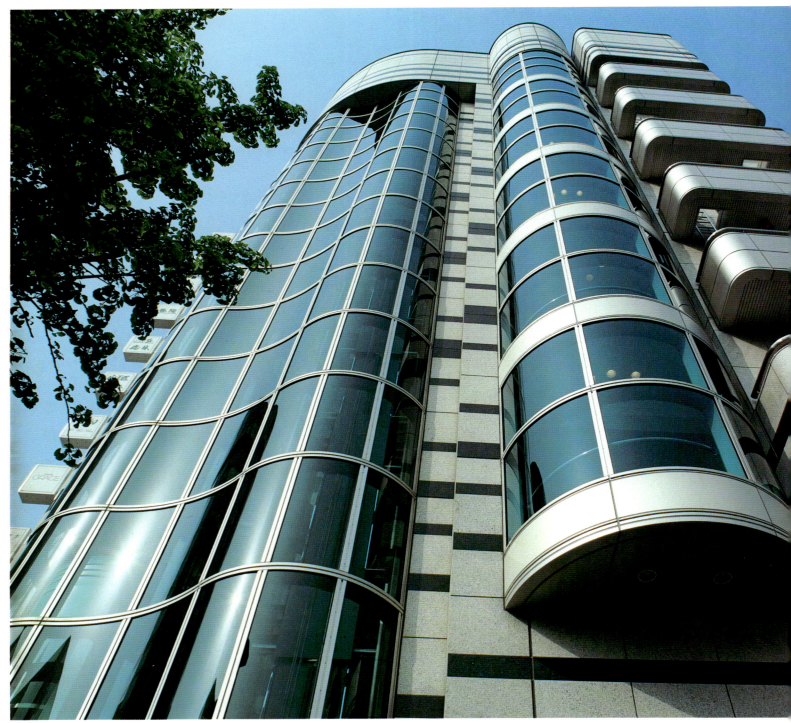

a：スパリゾートハワイアンズ・スプリングパーク
b：福島県いわき市
c：常磐興産
d：日本設計
e：日本設計，プロダクト翔/北川寿雄，稲垣ランドスケープデザイン研究所/稲垣丈夫
f：大成建設，常磐開発 J.V.
g：常磐興産/菅野哲也（菅野哲也写真事務所）
h：機能浴槽を隠しつつ，「自然の様」の創造（屋外）と驚きと発見の散策空間（屋内）。光ファイバー演出照明。
i：屋外：鮫川石，豆砂利，疑岩（FRP）/屋内：磁器タイル，疑岩，GRC，テフロン膜
●
j：日本設計

a : SPA RESORT HAWAIIANS SPRING PARK
b : Iwaki-shi, Fukushima
c : Joban Kosan Co.,Ltd.
d : Nihon Sekkei, Inc.
e : Nihon Sekkei, Inc.Hisao Kitagawa/Product SHOW Co.,Ltd.Takeo Inagaki/Inagaki Landscape Design
f : Taisei Corporation, Jobankaihatsu Co.,Ltd. J.V.
g : Joban Kosan Co.,Ltd./Tetsuya Kanno (Tetsuya Kanno Photo Office Co., Inc.)
h : SHIZEN NO SAMA (A state of nature) having a functional (bathtub outside) and a walking space which is full of amazement and discovery inside is created.
i : Exterior : Samekawa stone, gravel, cast stone(FRP)/Interior : Vitreous tile, cast stone, GRC, teflon film
j : Nihon Sekkei, Inc.

a：室津海水浴場整備
b：兵庫県津名郡
c：北淡町
f：志田建設興業
g：上谷重男
h：淡路島は今，リゾート開発のなかで風景美を失いつつある。それにインパクトを与えるデザインを。
i：鉄筋コンクリート平屋建
j：上谷重男＋U建築企画室
●

a：MUROZU
b：Tuna-gun, Hyogo
c：Hokutan Cho
f：Shida
g：Shigeo Uetani
h：In Awaji island, scenic beauty is fading away in the midst of the resort development. This is designed to give an impact upon the fading scenic beauty.
i：Reinforced concrete construction
j：Shigeo Uetani＋U Architectural Office

a：西神そごう出店計画
b：兵庫県神戸市
c：西神そごう
d：萬川純一，市川元則，天川雅晴
e：西山慎一郎，永松照基，吉村佳哲
f：アルス
h：「シンプル＆ソフト」というビジアュルコンセプトに基づき，洗練された感度の高い売場環境計画。
i：木地染色クリアー仕上げ，スチール焼付塗装，銅プロ・金メッキ，アイボリー塗装，ガラス等
j：アルス
●

a：SEISHIN SOGO DEPT. STORE PLANNING
b：Kobe shi, Hyogo
c：Seishin Sogo Dept.Store
d：Junichi Mankawa, Motonori Ichikawa, Masaharu Amakawa
e：Shinichiro Nishiyama, Terumoto Nagamatsu, Yoshitetsu Yoshimura
f：Arusu Co.,Ltd.
h：Environment of sales corner is designed to be a polished and hypersensitive on the basis of visual concept saying SIMPLE & SOFT.
i：The grain of wood clear finished, baked steel coating, copper brass, gold plating, ivory painting, glass etc.
j：Arusu Co.,Ltd.

a：F.A.S.もとまち
b：愛知県豊田市
c：豊田石油
d：伴田浩（都市計画）
e：福地誠（都市計画）
f：神谷組工業
g：吉倉光輝
h：都市空間に求められる文化生活情報を，手軽にエンターテイメントな空間で提供する複合型商業施設。
i：吹き付けタイル，ボンデ鋼板 SOP 塗，鉄骨 SOP 塗
j：都市計画
●

a：F.A.S.MOTOMACHI
b：Toyota-shi, Aichi
c：Toyota Sekiyu Co.,Ltd.
d：Hiroshi Banda
e：Makoto Fukuchi
f：Kamiya-Gumi Kogyo Co.,Ltd.
g：Kouki Yoshikura
h：A compound commercial establishment offering cultural life information of which an urban space need at an entertaining space conveniently.
i：Sprayed tile, bonderized sheet iron SOP painting, structural steel SOP painting
j：Toshikeikaku Inc.

a：東武スーパープール
b：埼玉県南埼玉郡
c：東武鉄道
d：上野卓二
e：寺島振介，三木正
f：鹿島建設
g：溝内了介
h：ニューリッチ感覚をもったヤングアダルトを本計画
のテーマとし，明るくヘルシーなウォーターパーク。
i：自然石エポキシ樹脂塗装，セラミック樹脂塗装，
ILB舗装，ステンレス及びRCプール
j：鹿島建設
●

a：TOBU SUPER POOL
b：Minami-Saitama-gun, Saitama
c：Tobu Tetsudo Co.,Ltd.
d：Takuji Ueno
e：Shinsuke Terashima, Tadashi Miki
f：Kajima Corporationon
g：Ryosuke Mizouchi
h：A bright and healthy water park design
under the theme of youg adults with a new rich sense.
i：Natural stone with epoxy resin coat,
ceramic resin coat, ILB paving, stainless steel
and RC pool
j：Kajima Corporation

a：東京セサミプレイス
b：東京都秋川市
c：東京セサミプレイス
d：鹿島建設/阿部和信
e：鹿島建設/小野寺康夫, セサミジャパン, 東京セサミプレイス
f：鹿島建設
g：鹿島建設/溝内了介
h：アメリカのセサミプレイスと同じ思想で, ライセンスを承諾して作られた子供のためのテーマパーク。
i：ILB, カラーアスコン, ケヤキ, モミジ, シラカジ, 他
j：鹿島建設
●

a：TOKYO SESAME PLACE
b：Akikawa-shi, Tokyo
c：Tokyo Sesame Place Ltd.
d：Kajima Corporation/Kazunobu Abe
e：Kajima Corporation/Yasuo Onodera, Sesame Japan Inc., Tokyo Sesame Place Ltd.
f：Kajima Corporation
g：Kajima Corporation/Ryosuke Mizouchi
h：A theme park for children designed with permission under the same concept as America's Sesame Place.
i：ILB, colored asphalt-concrete, zelkova trees, maple, white oak, etc.
j：Kajima Corporation

a：エクシブ白浜
b：和歌山県西牟婁郡
c：リゾートトラスト
d：日建設計/近藤繁
e：日建設計/山本範人，柴田嘉夫
f：鹿島建設
g：稲住泰広
h：太平洋に向かって広がる庭園内に列柱でプール空間を強調し、プールはアユや貝をモチーフにした。
i：プールサイド/レンガ、プール床/モザイクタイル50角模様張り
j：日建設計
●

a : XIV SHIRAHAMA
b : Nishi-Muro-gun, Wakayama
c : Resort Trust, Inc.
d : Nikken Sekkei/Shigeru Kondo
e : Nikken Sekkei/Norito Yamamoto, Yoshio Shibata
f : Kajima Corporation
g : Yasuhiro Inazumi
h : A colonade has been erected within the garden spreading to wards the Pacific Ocean in order to emphasize the space of the swimming pool which was designed with the motifs of Ayu and Shellfish.
i : Pool sides/brick, pool bottom/50mm square mosaic tiles
j : Nikken Sekkei Ltd.

a：芙蓉ミュージカルシアター
b：大阪府大阪市
c：芙蓉グループ花の万博'90出展委員会事務局
e：大成建設設計本部／川崎昭夫，森昌子
f：大成，前田，西松，飛島，五洋，鉄建，東亜共同企業体
g：大成建設広報部
h：一体の大樹に支えられたガラスの塔へ至るプロローグは，祝祭空間と緑とが融合する新しい都市。
i：カラー波板鉄板，吹き付けタイル，熱線反射ガラス
j：大成建設設計本部
●
a : FUYO MUSICAL THEATER
b : Osaka-shi, Osaka
c : Fuyo Group Expo'90 Exhibition Committee Secretariat
e : Taisei Corporation Design and Proposal
Division, Akio kawasaki, Masako Mori
f : Taisei, Maeda, Nishimatsu, Tobishima,
Goyo, Tekken & Toa J.V.
g : Taisei Corporation
h : A prologue reaching a glass tower
supported by one big tree is a new CITY
harmonizing a festival space and green.
i : Coloring ripple steel plate, pneumatically
applied tile, neat-reflective glass
j : Taisei Corporation Design and Proposal Division

a：新鳥羽水族館
b：三重県鳥羽市
c：鳥羽水族館
e：大成建設名古屋支店設計部/輿石勲，小菅敏彦，
濱口倫壽
f：大成建設名古屋支店
g：後藤秀雄（センター・フォト）
h：世界の海を再現し体験できる空間を構成し，メイン
通路での混雑を解消するよう動線計画した。
i：弾性アクリル吹き付けタイル，長尺塩ビシート，化粧
石膏ボード
j：大成建設設計本部
●
a：NEW TOBE AQUARIUM
b：Toba-shi, Mie
c：Toba Aquarium Corporation
e：Taisei Corporation Nagoya Branch Design
Division, Isao Koshiishi, Toshihiko Kosuge,
Michitoshi Hamaguchi
f：Taisei Corporation Nagoya Branch
g：Hideo Gotoh（Center Photo）
h：It is designed to structure a space where
the sea in all the world is reproduced and
people can feet it actually.As to the main
passage, a flow plan is designed to dissolve
crowdedness so far.
i：Elasticity acrylic spraying tile, long polyvinyl
chloride membrane, decorated gypsum boards
j：Taisei Corporation Design and Proposal Division

a：多摩そごうシンボルゾーン「多摩の四季」
b：東京都多摩市
c：多摩そごう
d：中辻伸
e：金子清二，児島正剛，原田文夫，エンジニア/鈴木俊幸，向隆宏
f：乃村工藝社
g：大東正巳
h：自然と人と文化の豊かな調和を願い、四季折々の多摩の自然の風景を野鳥たちの歌声にのせて再現。
j：乃村工藝社

●

a: TAMA SOGO TAMA-NO-SHIKI
b: Tama-shi, Tokyo
c: Tama Sogo Co.,Ltd.
d: Shin Nakatsuji
e: Seiji Kaneko, Masatake Kojima, Fumio Harada, Engineer/Toshiyuki Suzuki, Takahiro Mukai
f: Nomura Co.,Ltd.
g: Masami Daito
h: This area has been created by responding to the natural views afforded by the Tama area through each season a which resound with the chirping of the birds in express a wish for blessed harmony between nature, humans and culture.
j: Nomura Co.,Ltd.

85

a：銀座松屋リファイン120食品
b：東京都中央区
c：松屋
d：ゼナラルディレクター/内田繁，ディレクター/松川澄夫
e：荻野久男，福島勇人，皆神憲雄，市川浩一，細田理枝子，大達智子
f：乃村工藝社，松美舎
g：佐藤喜之
h：奥に細長い建築的デメリットを，シンプルな導線計画と適度に変化する環境デザインでメリットにした。
i：タイル，大理石，ナラ材，セラトーン
j：乃村工藝社
●
a: MATSUYA GINZA REFINE 120 FOOD
b: Chuo-ku, Tokyo
c: Matsuya
d: General Director/Shigeru Uchida, Director/Sumio Matsukawa
e: Hisao Ogino, Hayato Fukushima, Norio Minakami, Koichi Ichikawa, Rieko Hosoda, Tomoko Odachi
f: Nomura Co.,Ltd., Shobisya
g: Yoshiyuki Sato
h: The demerit of a deeply shaped construction has been turned into a merit by the simple flow-line plan and a moderate variation in the enviromental design.
i: Tiles, marble stone, oak timber, ceratone
j: Nomura Co.,Ltd.

Chapter 3

公共・オフィス空間デザイン
Public & Office Environmental Design

地域と共存する次世代のインテリジェント・オフィスビル
NEC スーパータワー

日本電気株式会社

21世紀を目前に控えた90年代のオフィス，新たな進化を求めた一つのこたえ。
それは周辺環境との融和のなかで，ひとりひとりが個性を十二分に発揮しながら
創造的な仕事に取り組めるビジネスアメニティーの実現である。
単に仕事の場としてでなく，生活の場としても，快適な空間のアメニティーに加え，
必要なときに必要な情報をフルに活用できる情報のアメニティーを装備した

創造的な次世代のインテリジェント・オフィスビル環境の提案といえる。
日本電気本社ビル「NEC スーパータワー」は，昨年，日本電気の創立90周年
を記念して建設された。NEC は1899年会社発祥のこの地で，地域社会との
協調，理解のもとに発展してきた歴史を念頭におき，新オフィスビル建設の検討
を長年重ねてきた。

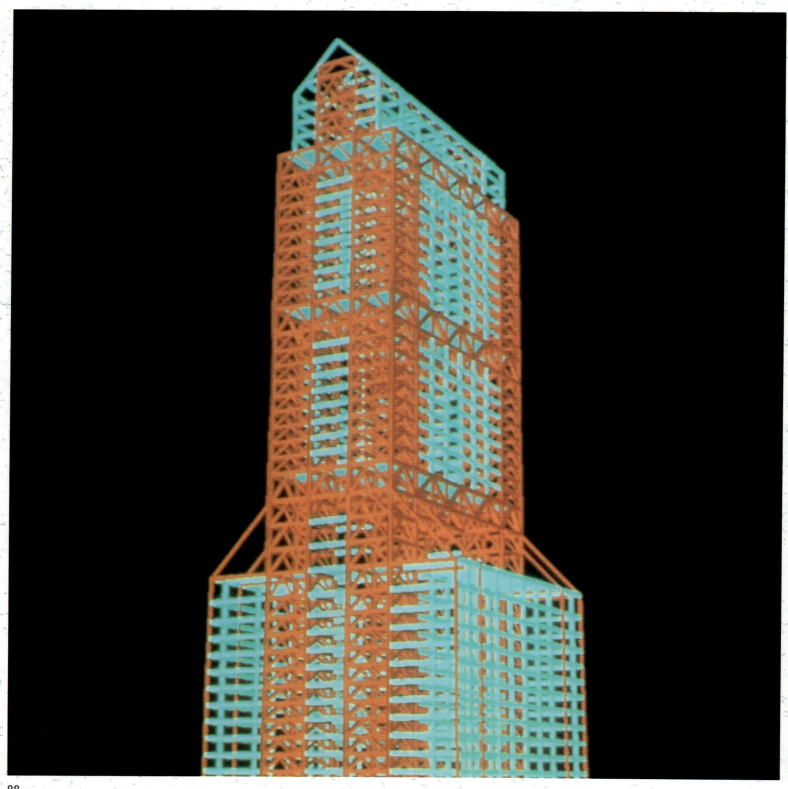

上にいくにしたがって段階的に細くなるスペースシャトルのような形状，
ビルの中ほどに開けられた巨大な風の通り道「ウインドアベニュー」，12階の高さ
まである大規模なアトリウム，敷地の3分の2以上を占める公園緑地など，
地域との環境調和を十分配慮した新しいオフィスビルのあり方を表現している。
超高層ビルで問題になるのはビル風と周辺の日照。ビルにぶつかった風は
ビルの壁面に沿って下へ向かい，地上に達して強い風になる。これを解決する
ために，13階から15階に相当する位置に風の通り道「ウインドアベニュー」を設け，
南北の風をそのまま通過させたり，ビルの形状に段差をつけ，風が地上に
達するまでの風圧を緩和し，同時に日照問題も最小限に抑える工夫が見られる。
ビルの全高は180m，地上43階，地下4階で，風の通り道「ウインドアベニュー」
はアトリウムへの自然採光窓ともなっている。
このビルは東京都と企業がよりよい街作りをするために定めた，東京都環境
影響評価条例の適用第1号の建築であり，それら環境問題，耐震性を解決する
ためには構造上の特長，スーパーフレームと呼ばれる強固な架構方式の
採用があって始めて果たせたのである（幅1m，厚さ100mmのH型鋼を
組み合わせた大組の柱4本を，約10階おきに巨大なトラス状鉄骨梁で連結した
架構方式で，ビル全体の加重をバランスよく支えている）。
ビジネス住環境を見てみよう。玄関ホールは高さ55mのアトリウム，この自然
採光のアトリウムを囲んで南北に配されたオフィスからは，社員がお互いの

仕事の場を見ることができ，このため社員の一体感が生まれる，いわばホロニック
な経営戦略を形で表したといえる。最上階にはVIP用レセプションホール，
高層階にはC&Cのハイテクノロジーで装備されたデシジョンルームやテレビ
会議室，CATVスタジオ，アトリウム最上階に面した2フロアは食堂，16階は
健康管理センター，フィットネスセンター，1，2階は接客ゾーン，サロン，
ガイダンスホール，地下1階は講堂と多目的ホールがある。ここには6000人の
社員が執務しているが，そのオフィスは最も進んだOA化オフィスであり，
最新のC&C（コンピュータと通信の融合）テクノロジーを結集，各フロアの
コンピュータとワークステーションは本社内だけでなく，衛星通信などによって
国内外の拠点と結ばれている。全世界NECグループ17万社員を結ぶ
ネットワーク「メッシュグローバリゼーション」を表現した伊藤隆道設計の巨大な
モニュメントが正面玄関前広場に設置されており，石井幹子デザインによる
ビルのライトアップに呼応し，季節に応じたライティングデコレーションで
夜のビジネス街に楽しさを提供している。
ひとりの知恵がみんなの知恵に。
このような考えでできあがったビジネスアメニティという新しい概念の
このインテリジェント・オフィスビルには，
来たるべき21世紀に向う示唆を見いだすことができる。

An intelligent office building for the next generation that will coexist with the local environment. NEC SUPER-TOWER

NEC Corporation

With the 21st century just around the corner, new lines of progress are sought for offices of the 90's. Business amenities which allow for more individuality and encourage creative work in harmony with the environment are necessary. These lines are suggestive of next generation creative intelligent office buildings which are equipped with amenities for supplying vast amounts of information when needed, as well as fulfilling the requirement of being pleasant and efficient places of work.

NEC Corporation's new headquarters, the NEC SUPER TOWER, was completed last year in celebration of the company's 90th anniversary. The building was constructed on the site where NEC first began operations in 1899. Every effort was made in the planning stage to make sure the comfort of local residents and office staff is taken into account in the every day running of the building. The eye is first struck by the unique shape of the building. Suggestive of a rocket, the SUPER TOWER narrows to an arrow-like point. Wind Avenue, a gigantic opening located at the center of the building, allows the free flow of wind through the structure. A large atruim occupies space from the ground to the 12th floor level.

Two-thirds of the site area is taken up by an artificial park popular with employees during lunch-time. These design features go towards creating a new office environment conscious of the vital need for harmony with the local surroundings.

One of the biggest problems posed by high-rise buildings to the immediate vicinity are high winds generated at ground level. Wind avenue considerably reduces WIND SHEAR by channelling strong wind through a gap between the 13th and 15th floors. The building is so positioned to capture and direct north-south winds predominant in the area away from the ground.

The NEC SUPER TOWER has 47 storeys (4 below ground) and rises to 180 metres. It is the seventh tallest building in Tokyo and the first building to fully comply with metropolitan environmental by-laws. The building is also well protected against the constant danger of earthquakes, for it employs an ultra-strong Super-Frame Structure which in tests has withstood earthquakes twice the magnitude of the Great Knato Earthquake of 1923.

Let's take a look at the business side of the environment. The 55m tall atrium serves as an entrance hall, and the offices situated to the north and south of this produce a feeling of oneness as the staff can see each other from their own office. The VIP reception room is on the top floor, the Decision Rooms TV conference rooms and CATV studio are all equipped with C&C high-technology and located on the upper floors. The two restaurant floors face the top floor of the atrium, the health control fitness center is on the 16th floor, the reception zone, saloon and guidance hall are on the first floor, and the main hall and multi-purpose halls are on the first basement level.

6,000 people work in each advanced OA office equipped with the latest C&C technology, and the computers and workstations on each floor are not only on-line within the building, but also with various places throughout the world by satelite link. In the square before the front entrance of the building rises a monument designed by Takamichi Ito expressing MESH GLOBALIZATION, NEC's corporate concept which links the 170,000 NEC workers in the NEC group throughout the world. In concert with the illumination of the building, designed by Motoko Ishii, the seasonal lighting decorations of this monument offer an exciting feel to the night in the Mita area. NEC's SUPER TOWER, pointing the way to the 21st century.

Floor Plans

Upper Levels

Middle Levels

Atrium

Lower Levels

Cross Section

Upper Floors 43F

38F

Middle Floors 27F

16F

Wind Avenue 12F

Lower Floors

Office   Office

Atrium

1F

Front Views

Side Views

43F レセプションホール 写真提供:毎日グラフ 撮影:山根敏郎

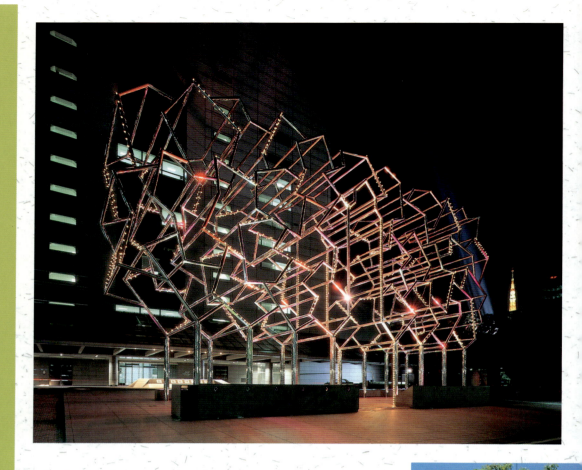

〈計画データ〉
建物名称＝日本電気本社ビル NEC スーパータワー
建築主＝日本電気
所在＝東京都港区芝5-7-1

〈設計〉
建築＝日建設計・東京/三栖邦博，岡本慶一，横田暉生，柴田好文，富樫亮
構造＝日建設計・東京/木原碩美，常木康弘
設備＝日建設計・東京/松縄堅，橋浦良介，野原文男，小林靖昌
造園＝日建設計・東京/森山明
インテリア＝日建設計・東京/吉川昭，大西芙紀
工務＝中村徹
協力＝風コンサルタント/風工学研究所，前庭モニュメント/伊藤隆道，外壁・照明デザイン/石井幹子
サインデザイン＝日本電気デザインセンター
監理＝日建設計・東京/辻愛也，石上智章，石川富夫，奥野正勝
デザインコーディネーション＝日本電気デザインセンター

〈施工〉
建築＝鹿島建設，大林組

〈構造〉
主体構造＝鉄筋造，鉄骨鉄筋コンクリート造，周辺地下/鉄筋コンクリート造
基礎＝高層部地下/ベタ基礎，周辺部地下/現場築造杭
階数＝地下4階，地上43階，塔屋1階建
設計期間＝1982年6月〜1986年8月
施工期間＝1986年11月〜1990年1月（既存建家，基礎解体撤去を含む）
敷地面積＝21,280平方メートル
建築面積・建ぺい率＝6,370平方メートル，30%
延床面積・容積率＝145,100平方メートル，600%

〈各階床面積〉
低層部基準階＝3,840平方メートル
中層部基準階＝2,450平方メートル
高層部基準階＝1,680平方メートル
地域地区＝商業地域，防火地区，特定街区指定
特殊な敷地環境＝東京都環境アセスメント
道路幅員＝東/27.0m，西/9.2m，南/10.0m，北/7.5m
階高・天井高＝低層部3.85m，中層部3.7m，CH2.6m
最高軒高・最高高さ＝178.4m，180.0m

〈外構〉
植栽＝クス，ケヤキ，ヤマモモなど
舗床＝磁器質タイル一部花こう岩 J&P，インターロッキングブロック

〈外部仕上げ〉
屋根＝アスファルト防水コンクリート押さえ，アトリウム屋根/パイプトラス，アルミサッシ，網入りガラス
外壁＝アルミパネル，フッ素樹脂塗装一部花こう岩 J&P
建具＝アルミ建具，フッ素樹脂塗装，2次電解着色

〈内部仕上げ〉
事務所＝床/コンクリート直均しタイルカーペット，壁/プラスターボード VE，天井/岩綿吸音板
1階サロン＝床/花こう岩 J&P，壁/花こう岩 J&P（乾式）一部アルミパネル，フッ素樹脂塗装，天井/アルミ2次電解着色鏡面仕上げ
食堂＝床/タイルカーペット一部パーケットフロア，壁/プラスターボード VE 一部結晶化ガラス，天井/岩綿吸音板，プラスターボード VE
役員応接室＝床/タイルカーペット，壁/ケイカル板下地クロス張り，天井/プラスターボード VE，岩綿吸音板
レセプションホール＝床/鋼製根太床，ケイ酸カルシウム板下地寄木張り（桜），壁/大理石（乾式），一部プラスターボード，ガラスクロス VE，天井/リブ付き岩綿吸音板

<Planning Data>
Project=NEC Head Office Building - NEC SUPER-TOWER
Client=NEC Corporation
Site=5-7-1, Shiba, Minato-ku, Tokyo

<Design>
Architecture=Nikken Sekkei Ltd., Tokyo/Kunihiro Misu, Keiichi Okamoto, Teruo Yokota, Yoshibumi Shibata, Ryo Togashi
Construction=Nikken Sekkei Ltd., Tokyo/Hiromi Kihara, Yasuhiro Tsuneki
Equipment=Nikken Sekkei Ltd., Tokyo/Katashi Matsunawa, Ryosuke Hashiura, Fumio Nohara, Yasumasa Kobayashi
Landscaping=Nikken Sekkei Ltd., Tokyo/Akira Moriyama

Interior=Nikken Sekkei Ltd., Tokyo/Akira Yoshikawa, Fuki Onishi
Engineering=Toru Nakamura
Cooperator=Wind Consultant/Wind Engineering Laboratory, Garden Monument/Takamichi Ito, External walls & Lighting/Motoko Ishii
Director/Yoshiya Tsuji, Tomoaki Ishigami, Tomio Ishikawa, Masakatsu Okuno

<Execution>
Contractor=Kajima Corporation, Ohbayashi Corporation

<Structure>
Main structure=Reinforced structure, reinforced steelframe concrete structure, surrounding basement/reinforced concrete.
Foundations=Multi-floored basement/mat foundations, surrounding basement/cast in place structure stakes.
Number of stories=4-level basement, 43 stories, 1 story tower
Design period=June 1982-August 1986
Construction period=November 1986-January 1990 (includes the demolition of former buildings)
Site area=21,280 sq.m.
Building area/building coverage=6,370 sq. m, 30%
Total floor area/rate of building volume=145, 100 sq.m, 600%

<Each floor area>
Lower level standard=3,840 sq.m
Middle level standard=2,450 sq.m
Higher level standard=1,680 sq.m
Area/district=Commercial area, fire zone, specified road
Special site environment=Tokyo municipal environment assessment
Width of road=East-27.0m, West-9.2m, South-10.0m, North-7.5m
Floor height/ceiling height=Lower part 3. 85m, middle part 3.7m, CH 2.6m
Maximum eaves height/maximum height= 178.4m, 180.0m

<Exterior Site>
Plantation=Camphor trees, zelcover, mountain peach trees, etc.
Paving=Ceramic tiles, part-granite J&P, inter-locking blocks

<Exterior Finish>
Roof=Asphalt waterproof covering, atrium roofing/pipe truss, aluminium sashes, wire reinforced glass
External walls=Aluminium panels, flouric resin coated partgranite veneer J&P
Fixtures=Aluminium fixtures, flouric resin coated, secondary electrolytic coloring

<Interior Finish>
Offices=floor/exposed concrete tiled carpet, walls/plaster board VE, Ceiling/rock wool acoustic boards
1st floor salon=Floor/granite veneer J&P, walls/granite veneer J&P (dry style), flouric resin coated part-aluminium panels, ceiling/ secondary electrolytic coloring mirror-surface finish aluminium
Dining room=Floor/tile carpet and part parquet, walls/plaster board VE, partly-crystalized glass, ceiling/plaster board VE, partly-crystalized glass, ceiling/rock wool acoustic boards and plasterboard VE
Reception room for board members=Floor/ tile carpet, walls/keikaru wood sheathed clother covering, ceiling/plaster board VE, rocj wool acoustic boards
Reception hall=Floor/steel joist flooring, calcium silicate wood sheathed parquet flooring (cherry wood), walls/large marble blocks (dry style), part plaster board, plass cloth VE, ceiling/ribbed rock wool acoustic boards

a：毎日放送本社ビル
b：大阪府大阪市
c：毎日放送
d：日建設計／上田信也
e：日建設計／中井進
f：大林組・竹中工務店 J.V.
g：東出清彦
h：M 型の特徴的な形態をもち，また都市を映しだす
ガラスカーテンウオールの開かれた都市型放送局。
i：ガラス，アルミ，花こう岩，ステンレス
j：日建設計
● 日建設計

a : MAINICHI BROADCASTING SYSTEM
HEADQUARTERS
b : Osaka-shi, Osaka
c : Mainichi Broadcasting System, Inc.
d : Nikken Sekkei/Shinya Ueda
e : Nikken Sekkei/Susumu Nakai
f : Obayashi gumi & Takenaka Corporation J.V.
g : Kiyohiko Higashide
h : This is an urban style broadcasting station
building characterized by its 'M' shape and
which is covered with a glass curtain to reflect
back the images of urban life.
i : Glass, aluminium, granite veneer, stainless steel
j : Nikken Sekkei Ltd.

a：クリスタルタワー
b：大阪府大阪市
c：竹中工務店，朝日ビルディング
e：竹中工務店設計部/山元弘和，本多友常
f：竹中工務店
g：庄野啓
h：クリスタルタワーはOBP地区を東西に走る片町徳庵線の西端に位置し，静かに発光する水晶体として想定された。
i：アルミ押出型材カーテンウォールポリウレタン樹脂塗装，熱線反射ガラス
j：竹中工務店
●
a：CRYSTAL TOWER
b：Osaka-shi, Osaka
c：Takenaka Corporation, Asahi Building Co., Ltd.
e：Takenaka Corporation, Building Design Dep./hirokazu Yamamoto, Tomotsune Honda
f：Takenaka Corporation
g：Hiraku Shono
h：this crystal tower located in the weaternmost part of the Katamachi Tokoan Line, which runs though the OBP from east to west, was designed to conjour up an image of a crystal body emitting gentle light.
i：Aluminium tube curtain awlls polyurethane resin coat, heat-resisting glass
j：Takenaka Corporation

a：フジタカ本社社屋
b：京都府長岡京市
c：フジタカ
d：クリエイティブディレクター＝森田耕二／アートディレクター＝岡田露愁
e：岡田露愁（アーティスト）
f：藤本工務店
g：森下友二
h：フジタカ本社社屋はルート171号線にあり、国道，名神高速道路，新幹線からもよく見える。
i：モニュメント：FRP
j：シルフ

●

a : FUJITACA CO., HEAD OFFICE BUILDING
b : Nagaokakyo-shi, Kyoto
c : Fujitaca Co.
d : Koji Morita, Roshiw Okada
e : Roshiw Okada
f : Fujimoto Komuten
g : Yuji morishita
h : Fujitaka Co., head office building is seen clearly from a national road, Meishin expressway and Shinkansen since it is located on Route 171.It creates a sensation around the circumference.
i : Monument : FRP
j : SYLPH

a：信用組合大阪興銀八尾支店
b：大阪府八尾市
c：信用組合大阪興銀
e：大成建設大阪支店設計部/美濃吉昭，内海利彦
f：大成建設
g：三沢ユタカ（イエローフラッグ）
h：多目的の最上階ホール天井は鉄骨単管梁のテフロン膜。昼夜で光の出入を反転させる。
i：四フッ化エチレン被覆膜，磁器タイル，ステンレス，曲面ガラス，スチールパネル
j：大成建設設計本部

●

a：UNION CREDIT OSAKAKOGIN YAO BRANCH
b：Yao-shi, Osaka
c：Union Credit Osakakogin
e：Taisei Corporation Design and Proposal Division, Yoshiaki Mino, Toshihiko Utsumi
f：Taisei Corporation
g：Yutaka Misawa
h：A ceiling of the topmost floor for multi-purpose is made of teflon filmed structural steel single tube beam.Going in and out of lights is reversed depending on which nighttime or daytime.
i：Tetrafluoride ethylene covering film, vitreous tile, stainless, curved surface glass, steel panel
j：Taisei Corporation Design and Proposal Division

100

a：福武書店本社ビル
b：岡山県岡山市
c：福武書店
d：日建設計/橋本巌
e：日建設計/寺岡俊彦
f：鹿島建設・大本組 J.V.
g：西日本写房
h：サンクガーデンをかいしてロビーと地下空間を立体的，開放的に外部環境と一体化した開かれた本社ビル。
i：ロビー床：花崗石，サンクガーデン床：板張り（南洋材）
j：日建設計

●
a：FUKUTAKE PUBLISHING BUILDING
b：Okayama-shi, Okayama
c：Fukutake Publishing Co.,Ltd.
d：Nikken Sekkei/Iwao Hashimoto
e：Nikken Sekkei/Toshihiko Teraoka
f：Kajima Corporation & Omoto-gumi, J.V.
g：Nishinihon Shabo
h：An open building of which the lobby and underground spaces have been integrated with the outside surroundings three-dimensionally and openly with a sunken garden as a border.
i：Lobby floor : granite veneer, sunken garden surface, planking (South Sea material)
j：Nikken Sekkei Ltd.

a：スタンフォード日本センター
b：京都府京都市
c：東洋信託銀行
e：三谷幸司，広田和正
f：大林組
g：島修一
h：風致地区の環境になじませるために玄関ホールを
スキップさせ，建物を半分地下に沈めている。
i：アスファルトシングル，アクリルリシン
j：大林組本店一級建築士事務所
●

a：STANFORD JAPAN CENTER
b：Kyoto-shi, Kyoto
c：Toyo Trust and Banking Co.,Ltd.
e：Koji Mitani, Kazumasa Hirota
f：Obayashi Corporation
g：Shuichi Shima
h：In order to a building integrate into
environment of a scenic zone, the entrance hall
is skipped and the building is half sunken.
i：Asphalt single, acrylic lysine
j：Obayashi Corporation

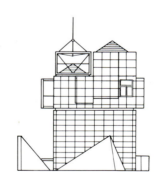

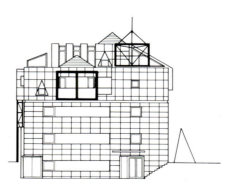

a：青山 M ビル
b：東京都港区
c：村田和夫
d：西澤健，南和正
e：松永博巳，白浜力，藤田雅俊
f：竹中工務店
g：T.ナカサ・アンド・パートナーズ
h：幾何学的な形態と色彩の建物は，雑然とした裏通りを行く人々に対して心地よい緊張感と新しい都市環境を作り出している。
i：御影石，アルミパネル，ステンレス，コンクリート打っ放し
j：GK 設計

●
a：AOYAMA M BUILDING
b：Minato-ku, Tokyo
c：Kazuo Murata
d：Takeshi Nishizawa, Masakazu Minami
e：Hiromi Matsunaga, Tsutomu Shirahama, Masatoshi Fujita
f：Takenaka Corporation
g：T.Nacasa & Partners
h：This gyrometric-shaped and colored building creates a comfortable tenseness and new city surrounding for the poeple who work alongside this back street.
i：Granite stone, aluminium-panels, stainless steel, exposed concerete
j：GK Sekkei Inc.

a：京都リサーチパーク
b：京都府京都市
c：京都リサーチパーク
d：日建設計/橋本巌
e：日建設計/矢島峰雄
f：大林組，鴻池組 他
g：SS 大阪
h：建物のロビーと連続した明るい中庭。自然素材と野草を中心とし，四季を折り込んだ床パターン。
i：花こう岩他自然石，土色の大型磁器質タイル，各種の豆砂利コンクリート洗い出し
j：日建設計
●

a : KYOTO RESEARCH PARK
b : Kyoto-shi, Kyoto
c : Kyoto Research Park Corporation
d : Nikken Sekkei/Iwao Hashimoto
e : Nikken Sekkei/Mineo Yajima
f : Obayashi-gumi, Konoike-gumi, etc.
g : SS Osaka
h : A bright courtyard connected to the lobby of the building.The four seasons have been integrated into the floor pattern design by using natural materials and wild grass patterns.
i : Granite veneer, natural stone, large earnthenware-colored ceamic tiles, various types of washed gravel
j : Nikken Sekkei Ltd.

a：大同生命福井第2ビル
b：福井県福井市
c：大同生命保険相互会社
e：藤縄正俊，船橋昭三，星野一
f：大林組，熊谷組共同企業体
g：スタディオ クリオ，上諸尚美
h：外観デザインに特徴をもたせVカットしたガラス面で，街行く人を受け止め店舗へ入りやすい形とした。
i：花こう岩（本磨），ミラーガラス
j：大林組本店一級建築士事務所
●

a : DAIDO SEIMEI FUKUI # 2 BLDG.
b : Fukui-shi, Fukui
c : Daido Mutual Life Insurance Company
e : Masatoshi Fujinawa, Shozo Funahashi, Hajime Hoshino
f : Obayashi Corp., Kumagai Co.,Ltd. J.V.
g : Sutadio Crio Naomi Kamimoro
h : A shop is designed to catch walkers and make them feel willing to enter thereby V-cutted-glass wall constituting a characteristic feature of external ap-pearance design.
i : Granite(polished finish), mirror glass
j : Obayashi Corporation

a：松下IMPビル
b：大阪府大阪市
c：松下アイ・エム・ピー
d：日建設計/寺本敏則
e：日建設計/小野為一，菊田靖久
f：鹿島建設・大林組・熊谷組・戸田建設・銭高組・前
田建設工業・松村組 J.V.
g：稲住泰広，柄松稔
h：大阪ビジネスパークに立地した，200メートルにおよ
ぶショッピングモールとオフィスの複合ビル。
i：外装/ラスタータイル模様張り，内装・モール床/大
理石模様張り
j：日建設計
●
a：MATSUSHITA IMP BUILDING
b：Osaka-shi, Osaka
c：Matsushita IMP Co, Ltd.
d：Nikken Sekkei/Toshinori Teramoto
e：Nikken Sekkei/Tameichi Ono, Yasuhisa Kikuta
f：Kajima, Obayashi-gumi, Kumagaya-gumi,
Toda Kensetsu, Zenidaka-gumi, Maeda
Kensetsu and Matsumura-gumi, J.V.
g：Yasuhiro Inazumi, Minoru Karamatsu
h：A complex building containing a shopping
mall that stretched for 200m and offices
located in the middle of this business park in Osaka.
i：Exterior decoration/lustre-patterned tiling,
interior decoration/mall floor/marble stone,
patterned tiling
j：Nikken Sekkei Ltd.

a：ニッセイ・ライフ・プラザ新宿（新宿NSビル1階）
b：東京都新宿区
c：日本生命保険相互会社
d：藤井栄一
e：オブジェ家具デザイン/ヴィンチェンツォ・イアヴィコリ，マリア・ルイザ・ロッシ，絵画/高柴牧子
f：大成建設
g：大東正巳
h：若者の生活の夢づくりにアトリウムに面した親しみを，オブジェ家具によって表現した情報プラザ。
i：タイルカーペット，スタッコ，パンチングメタル，ア ルミルーバー，メープル材
j：日建設計
●
a：NISSAY LIFE PLAZA SHINJUKU
b：Shinjuku-ku, Tokyo
c：Nippon Life Insurance Company
d：Eichi Fujii
e：Object/Vincenzo Iavicoli, Maria Luisa Rossi,
Art/Makiko Takashiba
f：Taisei Corporation
g：Masami Daito
h：The friendliness of this information plaza
facing the atrium has been expressed by the
use of object furniture in order to promote the
dream-like lifestyles of young poeple.
i：Tile carpetting, stucco, punched metal, aluminium louvres, maple
j：Nikken Sekkei Ltd.

a：ニッセイ高知本町ビル
b：高知県高知市
c：日本生命保険相互会社
e：田中弘成，本間利雄，山田宗彦
f：大林組・大旺建設・轟組共同企業体
g：花城辰男（SS 大阪）
h：ブルーとグレーのミラーガラスに，南国の青空と高
知城下の並木を映して緑豊かな環境を引立てた。
i：高性能熱線反射ガラス，磁器質タイル，外国産花こう石
j：大林組本店一級建築士事務所
●
a：NISSAY KOCHI HOMMACHI BUILDING
b：Kochi-shi, Kochi
c：Nippon Life Insurance Company
e：Hironari Tanaka, Toshio Honma, Kazuhiko Yamada
f：Obayashi, Daio, Todoroki J.V.
g：Tatsuo Hanashiro (SS Osaka)
h：A building is designed to emphasize the
rich green environment by mirror glasses
colored blue and gray reflecting blue sky of
the South and tree lines at the foot of Kochi castle.
i：High-performance heat-reflective glass,
vitreous tiles, imported granite
j：Obayashi Corporation

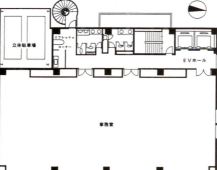

a：東京生命本社ビル
b：東京都千代田区
c：東京生命保険相互会社
e：大成建設設計本部/佐野隆夫，秋山哲也，山田達行
f：大成建設東京支店
g：SS東京
h：サンクンガーデンの緑を斜めに掛けられたミラーガラスに映し出し，街に緑の光を投げかける。
i：花こう岩，ハーフミラーガラス，アルミ
j：大成建設設計本部
●
a：HEAD OFFICE TOKYO LIFE INSURANCE
COMPANY BUILDING
b：Chiyoda-ku, Tokyo
c：Tokyo Life Insurance Company
e：Taisei Corporation Design and Proposal
Division, Takao Sano, Tetsuya Akiyama,
Tatsuyuki Yamada
f：Taisei Corporation Tokyo Branch
g：SS Tokyo
h：Sunkun garden's green is reflected by a
mirror glass hung obliquely and glows of
green through at the city.
i：Granite, half mirror glass, aluminum
j：Taisei Corporation Design and Proposal Division

a：大同生命山形ビル
b：山形県山形市
c：大同生命保険相互会社
e：大成建設設計本部，山口一平，塚田哲也
f：大成建設
g：菅野哲也
h：「冬の山形にあって存在感のあるオフィスビルを」がコンセプト。南北面外壁は半径250mの曲面壁となっている。
i：外壁：二丁掛ラスタータイル/柱型：アルミパネル/ホール壁：大理石
j：大成建設設計本部

●

a: DAIDO LIFE YAMAGATA BUILDING
b: Yamagata-shi, Yamagata
c: Daido Life Insurance Company
e: Taisei Corporation Design and Proposal Division, Kunihira Yamaguchi, Tetsuya Tsukada
f: Taisei Corporation
g: Tetsuya Kanno
h: Its concept is An outstanding office building in Yamagata in winter. South and north exterior wall faces are curved surface walls with a radius of 250m.
i: Exterior wall : Double sized luster tile/ Columns : Aluminum panel/Hall wall : Marble
j: Taisei Corporation Design and Proposal Division

a：NKF 川崎ビルディング
b：神奈川県川崎市
c：日本鋼管不動産
e：竹中工務店設計部/宮田勝弘，田代廣信，松岡俊之
f：竹中工務店
g：斎部巧
h：端正な表情のなかに，親しみやすさとヒューマンな感覚をかねそなえたオフィスビル。
i：外壁：黒御影石本磨き/エントランス：大理石本磨き/オブジェ：スチールパネル焼付塗装
j：竹中工務店
●
a：NKF KAWASAKI BUILDING
b：Kawasaki-shi, Kanagawa
c：NKF Corporation
e：Takenaka Corporation, Building Design Dep./katsuhiro Miyata, Hironobu Tashiro, Toshiyuki Matsuoka
f：Takenaka Corporation
g：Isao Imbe
h：This office building combinees a friendly atmosphere and a sense of humanity in its classical face.
i：Black granite with a polished finish (exterior wall), polished marble (entrance), enamel coated steel panelling
j：Takenaka Corporation

a：チチヤス本社ビル
b：広島県佐伯市
c：チチヤス乳業
e：竹中工務店設計部／井上軍，小林泰樹
f：竹中工務店
g：国際写真
h：自然の中のオフィスをキーワードに，恵まれた自然
環境との交流と調和をテーマとした。
i：外壁：二丁掛磁器タイル，軽量コンクリートパネル／
サッシュ：アルミ
j：竹中工務店
●
a：CHICHIYASU HEAD OFFICE
b：Saeki-shi, Hiroshima
c：Chichiyasu Milk Products Co.,Ltd.
e：Takenaka Corporation, Building Design
Dep./Isao Inoue, Yasuki Kobayashi
f：Takenaka Corporation
g：Kokusai Shashin
h：Using「An Office in the Midst of Nature」as a
keyword, the major theme here was to offer
harmonious exchange with the surrounding
nature.
i：Exterior wall : Double-ply ceramic tilles,
lightweight concrete panel/Sashes : aluminium
j：Takenaka Corporation

a：大阪南 YMCA 会館
b：大阪府大阪市
c：大阪キリスト教青年会
e：竹中工務店設計部/本城邦彦，前田俊雄
f：竹中工務店
g：庄野啓
h：塔状の渡り廊下のある中庭を囲んで建物を配置し，
外部・内部空間相互の貫入を図った。
i：スチール亜鉛メッキ仕上げ，アルミ，木，天然リノリウム
j：竹中工務店

●

a：OSAKA MINAMI YMCA BUILDING
b：Osaka-shi, Osaka
c：Osaka YMCA
e：Takenaka Corporation, Building Design
Dep./Kunihiko Honjo, Toshio Maneda
f：Takenaka Corporation
g：Hiraku Shono
h：This building has been arranged around a
courtyard which has a tower-shaped
connecting corridor in order to create a
passageway between the outside and inside
i：Zinc plated steel, aluminium, wood, natural linolium
j：Takenaka Corporation

a：NIKKI-SONY 共同ビル
b：東京都品川区
c：日本気化器製作所，ソニー
e：竹中工務店設計部/服部紀和，豊村博，地主道夫
f：竹中工務店
g：マツダ・プロ・カラー/吉村行雄
h：一日の，そして季節の移り変りをファサードに映し，新しいランドマークとなる。
i：ケヤキ，カツラ，熱線反射ガラス，磁器タイル，アルミ，花こう岩，大理石，ピンコロ石
j：竹中工務店
●

a：NIKKI-SONY BUILDING
b：Shinagawa-ku, Tokyo
c：Nippon Garbureter Co.,Ltd., SONY Corporation
e：Takenaka Corporation, Building Design Dep./Norikazu Hattori, Hiroshi Toyomura, Michio Jinushi
f：Takenaka Corporation
g：Matsuda Pro Color/Yukio Yoshimura
h：Reflecting the variations of the day and the seasons on its facade, this building will become a new landmark.
i：Katsura trees, heat-reflecting glass, ceramic tiles, aluminium, granite veneer, marble, pin stones
j：Takenaka Corporation

a：横浜ビジネスパーク
b：神奈川県横浜市
c：野村不動産
d：YBP 設計室（野村不動産＋大林組）
e：YBP 設計室（野村不動産＋大林組），牛山恭男，
毛塚洋，尾関勝之，川瀬俊二，井出昭治
f：大林組
g：白鳥美雄
h：横浜ビジネスパークは，民間最大級の脱都心型イ
ンテリジェントシティである。
i：花こう岩，大理石，アルミカーテンウォール，ステ
ンレス，ミッドランドブリック
j：YBP 設計室（野村不動産＋大林組）
●

a : YOKOHAMA BUSINESS PARK
b : Yokohama-shi, Kanagawa
c : Nomura Real Estate Development Co.,Ltd.
d : YBP Design Room（Nomura Real Estate
Development Co.,Ltd.+Obayashi Corporation）
e : YBP Design Room（Nomura Real Estate
Development Co.,Ltd.+Obayashi Corporation）:
Yasuo Ushiyama, Hiroshi Kezuka, Katsuyuki
Ozeki, Shunji Kawase, Shyoji Ide
f : Obayashi Corporation
g : Yoshio Shiratori
h : Yokohama Business Park is the one of the
biggest intelligent cities which have a
conception of being part from the center of
the Metropolis.
i : Granite, marble, aluminum curtain wall,
stainless, midland brick
j : YBP Design Room（Nomura Real Estate
Development Co.,Ltd.+Obayashi Corporation）

a：和歌山朝日ビル
b：和歌山県和歌山市
c：朝日ビルディング
e：竹中工務店設計部/柏木浩一，酒井利行
f：竹中工務店
g：松村芳治
h：城と対峙して建ち，新聞社の入るビルとして，昼夜を問わず街のランドマークとなることを目指した。
i：ガラスブロック，磁器タイル，コンクリート
j：竹中工務店

●
a：WAKAYAMA ASAHI BUILDING
b：Wakayama-shi, Wakayama
c：Asahi Building
e：Takenaka Corporation, Building Design Dep./Koichi Kashiwagi, Toshiyuki Sakai
f：Takenaka Corporation
g：Yoshiharu Matsumura
h：Standing face to face with the castle, this building has been designed to become a landmark on the street anytime of the day or night as a building that accomdates a press office.
i：Glass blocks, ceramic tiles, concrete
j：Takenaka Corporation

a：新日鉱ビル
b：東京都港区
c：虎ノ門タワービルディング
d：小池俊之
e：富松太基，柳学
f：鹿島建設，他 J.V.
g：川澄建築写真事務所
h：L字型の2棟のビルの間に高さ80m のアトリウムを
とり，人々の憩いの空間として親しまれている。
i：壁：アルミ，高性能熱線反射ガラス/床：御影石，
モニュメント：チタン（電解発色）
j：日本設計
●
a: Shin-Nikko Building
b: Minato-ku, Tokyo
c: Toranomon-Tower Building Co.
d: Toshiyuki Koike
e: Taiki Tomatsu, Gaku Yanagi
f: Kajima Corp., etc., J.V.
g: Kawasumi Architectural Photograph Office
h: An 80 m high atrium which are between two
of L-shaped buildings is familiarized.
i: Wall : Aluminum, high-performance heat-
reflective glass/Floor : Granite/Monument :
Titanium (Integral colour anodic oxidation)
j: Nihon Sekkei, Inc.

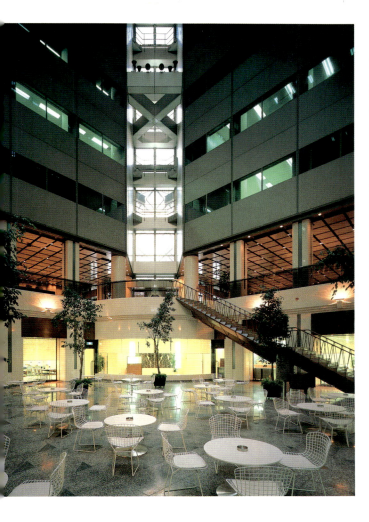

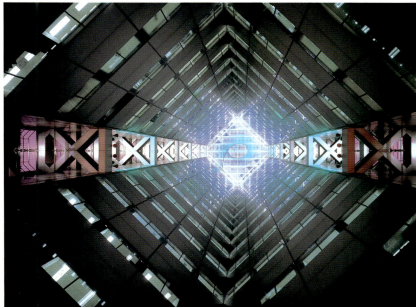

a：沖ビジネス本社ビル
b：東京都港区
c：沖ビジネス
d：沖ディベロップメント
e：大成建設設計本部/山口一平，河野晴彦
f：大成建設・大木建設共同企業体
g：野口毅
h：運河に向かって開かれた外観と商談コーナー，街角の吹抜とグリーン，屋上庭園をもつ社員クラブ。
i：外装：PC板45×二丁掛ラスタータイル，アルミサッシ横連窓線熱反射ガラス
j：大成建設設計本部

a : OKI BUSSINESS BUILDING
b : Minato-ku, Tokyo
c : Oki Business Co.,Ltd.
d : Oki Development Co.,Ltd.
e : Taisei Corporation Design and Proposal Division, Kunihira Yamaguchi, Haruhiko Kohno
f : Taisei, Oki J.V.
g : Takeshi Noguchi
h : Exterior and a business talk corner facing to a canal, an atrium and green on a street corner, a staff club room which has a roof garden.
i : Exterior : 45×95 sq mm sized PC panel with luster tile, heat reflecting glass for aluminum sash horizontal multiple window
j : Taisei Corporation Design and Proposal Division

a：新宿171ビル
b：東京都新宿区
c：浜屋ガラス
e：大成建設設計本部/山口一平，河野晴彦
f：大成建設
g：三輪晃士
h：映りこみ効果のカーテンウォールと，吹抜やテラス
の構成が建物の内外の表情に変化を与える。
i：外壁：アルミカーテンウォール，熱線反射ガラス，
一部アルミパネル，45×二丁掛ラスタータイル
j：大成建設設計本部
●
a：SHINJUKU 171 BUILDING
b：Shinjuku-ku, Tokyo
c：Hamaya Co., Ltd
e：Taisei Corporation Design and Proposal
Division, Kunihira Yamaguchi, Haruhiko Kohno
f：Taisei Corporation
g：Kohshi Miwa
h：Curtain wall having reflection effect and
structure of an atrium and terrace give the
buildings' internal and external expression variety.
i：Exterior：Aluminum curtain wall heat-
reflective glass partly aluminum 45 x 95 sq mm
sized panel with luster tile
j：Taisei Corporation Design and Proposal Division

a：新川崎三井ビルディング
b：神奈川県川崎市
c：三井不動産
f：鹿島建設，三井建設，三井不動産建設 J.V.
g：川澄建築写真事務所
h：郊外型オフィスのアメニティ空間としてアトリウムを
設置。ハイテックなデザインを目指した。
i：高性能熱線反射ガラス，アルミフッ素樹脂塗装，結晶化ガラス
j：日本設計
●
a：SHINKAWASAKI MITSUI BUILDING
b：Kawasaki-shi, Kanagawa
c：Mitsui Real Estate Development Co.,Ltd.
f：Kajima Corp., Mitsui Construction Co, Ltd.,
Mitsui Harbour and Urban Construction Co.,Ltd.J.V.
g：Kawasumi Architectural Photograph Office
h：An atrium is set as an amenity space of a
suburban office.Its aim is high-tech design.
i：High-performance heat-reflective glass,
aluminum fluorine resin painting, crystallization glass
j：Nihon Sekkei, Inc.

a：新橋 NH ビル
b：東京都港区
c：日本ヒューム管
e：大成建設設計本部/田辺詔二郎
f：大成建設東京支店
g：野口写真事務所
h：小規模ながらも前庭を設け，歩道空間と一体となって外観にゆとりが生まれるように配慮した。
i：磁器タイル，一部御影石の打込 PC 板
j：大成建設設計本部
●

a : SHINBASHI NH BLDG.
b : Minato-ku, Tokyo
c : Nippon Hume Pipe Co.,Ltd.
e : Taisei Corporation Design and Proposal Division, Shojiro Tanabe
f : Taisei Corporation Tokyo Branch
g : Noguchi Photo Studio
h : It is designed to express latitude of exterior by making a small scale entrance court uniting to a foot path space.
i : Vitreous tile and partly granite concreting, PC panels
j : Taisei Corporation Design and Proposal Division

a：白山オフィス
b：東京都文京区
c：丹青社
d：堀越裕二
e：堀越裕二
f：丹青社
g：米倉写真事務所
h：オフィス内を格子窓越しに見るような空間演出およびオープンなミーティングスペース。
i：メイプル，タイルカーペット，ブラインドガラス，スタッコアンティーコ
j：丹青社オフィスデザイン研究所
●

a : HAKUSAN OFFICE
b : Bunkyo-ku, Tokyo
c : Tanseisha Co.,Ltd.
d : Yuji Horikoshi
e : Yuji Horikoshi
f : Tanseisha Co.,Ltd.
g : Eiji Yonekura
h : Office space is produced as if inside of office is seen from a lane through a lattice window and opened meeting space.
i : Maple, tile carpet, blind glass, stucco antico
j : Tanseisha Co.,Ltd.Office Design Institute

a：アライアンス
b：東京都港区
c：岩佐凱實／青南ランド
e：大成建設設計本部／西条由治，桑名正一
f：大成建設
g：北沢治夫（SS東京）
h：ファッションの街，青山にふさわしいハイセンスなオフィスビル。
i：外国産花こう岩，磁器タイル，連窓アルミサッシ
j：大成建設設計本部
●

a：ALLIANCE
b：Minato-ku, Tokyo
c：Yoshizane Iwasa/Seinanrando Corporation
e：Taisei Corporation Design and Proposal Division, Yoshiharu Nishijo, Masaichi Kuwana
f：Taisei Corporation
g：Haruo Kitazawa (Corporation.SS.Tokyo)
h：A good sense office building which is suitable for a fashionable streets Aoyama.
i：Imported granite, vitreous tile, multiple window aluminum sash
j：Taisei Corporation Design and Proposal Division

a：106 GINZA BLDG.
b：東京都中央区
c：土志田合資会社／土志田眞次郎，井ノ下栄
e：大成建設設計本部／山口一平，橋本緑郎
f：大成建設
g：野口毅
h：銀座に建つ小規模オフィスビル。地下の光庭や最上階のボールトに工夫があり，印象的なデザインとなっている。
i：二丁掛リブ付ラスタータイル，花こう岩ジェットバーナー，ステンレスフラットバー
j：大成建設設計本部
●

a：106 GINZA BLDG.
b：Chuo-ku, Tokyo
c：Toshita Goshigaisha/Shinjiro Toshita, Sakae Inoishita
e：Taisei Corporation Design and Proposal Division, Kunihira Yamaguchi, Rokuro Hashimoto
f：Taisei Corporation
g：Takeshi Noguchi
h：A small-scale office building standing in Ginza.An underground luminous garden and top floor vault are contrived so that it may be an impressive design.
i：Double sized ribbed luster tile, granite jet burner, stainless flat bar
j：Taisei Corporation Design and Proposal Division

a：さぬき信用金庫本店
b：香川県丸亀市
c：さぬき信用金庫
e：大成建設四国支店設計部/伊藤治磨
f：大成建設四国支店
g：合田建築写真研究所
h：内部機能をそのまま外装デザインに取り組み，地方
都市に相応した信用金庫とした。
i：御影石，二丁掛タイル，アルミパネル
j：大成建設設計本部
●

a : CREDIT BANK OF SANUKI HEAD OFFICE
b : Marugame-shi, Kagawa
c : Credit Bank of Sanuki
e : Taisei Corporation Shikoku Branch Design
Division, Harumaro Ito
f : Taisei Corporation Shikoku Desigh
Division
g : Gouda Office
h : A credit bank is designed to fit a provincial
city by reflecting internal function into exterior
design.
i : Granite, double sized tile, aluminum panel
j : Taisei Corporation Design and Proposal
Division

a：小倉ビル
b：東京都中央区
c：小倉興業
d：番晶勝，東京建物
e：大成建設設計本部/山口一平，今里清，鈴木裕
美，佐溝裕紀，中村和久
f：大成建設
g：服部寿徳（ハットリスタジオ）
h：セットバックによる緑地などの環境整備。重量感の
あるファサード。地階の居室化。
i：御影石
j：大成建設設計本部
●

a : OGURA BUILDING
b : Cyuo-ku, Tokyo
c : Ogura Kogyo Corporation
d : Masakatsu Ban, Tokyo Tatemono
e : Taisei Corporation Design and Proposal
Division, Kunihira Yamaguchi, Kiyoshi Imasato,
Yumi Suzuki, Yasunori Samizo, Kazuhisa
Nakamura
f : Taisei Corporation
g : Toshinori Hattori (Hattori Studio)
h : A setback is utilized to keep a good
environment such as green area.A facade
which are massive and tidy.Basement floor is
created as a sitting room.
i : Granite
j : Taisei Corporation Design and Proposal
Division

a：新宿モノリス
b：東京都新宿区
c：安田信託銀行（代表受託者）
e：彫刻：澄川喜一，照明計画基本：石井幹子
f：間組，他
g：川澄建築写真事務所
h：公開空地である歩行者広場に対し，街路の延長となるようなアトリウム空間を提供したいと考えた。
i：壁：イタリア産大理石／床：国産御影石／ロビー天井：アルミパネル，フッ素樹脂塗装
j：日本設計
●

a : SHINJUKU MONOLITH
b : Shinjuku-Ku, Tokyo
c : The Yasuda Trust and Banking Company, Limited
e : Kiichi Sumikawa, Motoko Ishii
f : Hazama Gumi, Ltd.etc.
g : Kawasumi Architectural Photograph Office
h : It is designed to offer an atrium space which will be an extension of a street to a pedestrian's park which are an open place.
i : Wall : Italian marble/Floor : Japanese granite/Lobby ceiling : Aluminum panel, fluorine resin painting
j : Nihon Sekkei, Inc.

a：レランドセンタービル
b：千葉県船橋市
e：大成建設設計本部／山口一平，橋本緑郎
f：大成建設東京支店
g：狩野正和，野口毅
h：中庭をもったオフィスビル。1F，2Fまわりを公共に開放し，楽しげな演出をしている。
i：アルミパネル，タイル，花こう岩
j：大成建設設計本部
●

a : RELAND CENTER BUILDING
b : Funabashi-shi Chiba
e : Taisei Corporation Design and Proposal Division, Kunihira Yamaguchi, Rokuro Hashimoto
f : Taisei Corporation Tokyo Branch
g : Masakazu Kano, Takeshi Noguchi
h : An office building which has a courtyard.It is produced pleasantly by throwing open around 1 and 2 floor to the public.
i : Aluminum panel, tile, granite
j : Taisei Corporation Design and Proposal Division

a：武藤工業本社ビル
b：東京都世田谷区
c：武藤工業
e：竹中工務店設計部/矢口進，井上雅雄
f：竹中工務店
g：東京グラフィック
h：地下鉄コンコースと直結した屋内型公開空地とアトリウムを要に回遊する楽しさをもつオフィス。
i：ステンレスおよび鉛複合板横葺き屋根，群青色特殊磁器タイル，アトリウム大理石本磨
j：竹中工務店
●

a : HEAD-QUARTERS OF MUTO INDUSTRIES
b : Setagaya-ku, Tokyo
c : Muto Industries
e : Takenaka Corporation, Building Design Dep./Susumu Yaguchi, Masao Inoue
f : Takenaka Corporation
g : Tokyo Graphic
h : With its indoor-style public open space and atrium directly connected to the subway concourse, a circular tour can be enjoyed in this office.
i : Stainless steel and lead complex = plywood for the horizontal panelling, special ultramarine ceramic tiles, atrium marble with polished finish
j : Takenaka Corporation

a：備後町山口玄ビル
b：大阪府大阪市
c：山口玄洞
e：藤縄正俊，船橋昭三
f：大林組
g：上諸尚美
h：浮遊する最上階を支える柱，梁の格子組が，緑の公開空地に際立つよう心がけた。
i：花こう岩，アルミサッシ，熱線吸収・反射ガラス，公開空地の植樹
j：大林組本店一級建築士事務所
●

a : BINGO-MACHI YAMAGUCHIGEN BLDG.
b : Osaka-shi, Osaka
c : Gendo Yamaguchi
e : Masatoshi Fujinawa, Shozo Funahashi
f : Obayashi Corporation
g : Naomi Kamimoro
h : It is designed to be conspicuous a pillar and latticing of beam supporting the highest floor which is floated in an open space with greenery.
i : Granite, aluminum sash, heat absorbing glass, heat-reflective glass, greenery of an open space
j : Obayashi Corporation

a：TNN放送センター社屋（テレビネットワーク・ノベオカ）
b：宮崎県延岡市
c：テレビネットワーク延岡
d：小嶋凌衛
e：小嶋凌衛
f：西松・上田建設共同企業体
g：小嶋凌衛
h：情報量・スピード，ともにハンデを負う地方都市の情報発信基地として先端性を形態，環境造りで訴えた。
i：ホーロー MC ライト/半鏡面ガラスアルミカーテンウォール/ALC 板
j：小嶋凌衛建築設計事務所，小嶋凌衛

a：TV-NETWORK NOBEOKA
b：Nobeoka-shi, Miyazaki
c：TV-Network Nobeoka Co.,Ltd.
d：Ryoe Kojima
e：Ryoe Kojima
f：Nishimatsu, Ueda J.V.
g：Ryoe kojima
h：It is designed to make environment for an ultramodern information transmission base of provincial city which has handicaps on quantity and rapidity of information.
i：Hollow MC light, half mirror finished glass aluminum curtain wall, ALC board
j：Ryoe Kojima and Associates

a：神田ホリイビル
b：東京都千代田区
c：ホリイ/堀井仙太郎
e：大成建設設計本部/山口一平，橋本緑郎
f：大成建設
g：野口毅
h：狭い間口のオフィスビルで，最上階の切妻屋根と階高の高い1階が特徴となっている。
i：45二丁掛タイル，カラーステンレス屋根材，花こう岩ジェットバーナーおよび本磨き
j：大成建設設計本部
●

a：KANDA HORII BLDG.
b：Chiyoda-ku, Tokyo
c：Horii Corporation, Sentaro Horii
e：Taisei Corporation Design and Proposal Division, Kunihira Yamaguchi, Rokuro Hashimoto
f：Taisei Corporation
g：Takeshi Noguchi
h：A narrow frontage office building whose distinguishing marks are the gable roof on the topmost floor and the first floor which is high story height.
i：Double sized porcelain tiles 45 mm square, color stainless roof material, granite jet burner and polished
j：Taisei Corporation Design and Proposal Division

a：太陽生命高知ビル
b：高知県高知市
c：太陽生命保険相互会社
e：藤縄正俊，船橋昭三，広田和正，河南誠
f：大林組
g：竹村写真事務所
h：生保ビルに不可欠な広告塔を建物のデザインと一体化し，周辺環境から一際目立つようにした。
i：磁器質タイル，熱線吸収ガラス
j：大林組本店一級建築士事務所
●

a：TAIYO SEIMEI KOCHI BLDG.
b：Kochi-shi, Kochi
c：Taiyo Seimei
e：Masatoshi Fujinawa, Shozo Funahashi, Kazumasa Hirota, Makoto Kannan
f：Obayashi Corporation
g：Takemura Photo Office
h：It is designed to stand out clearly from surroundings by uniting an advertising pillar which is indispensable for a life insurance company's building.
i：Vitreous tile, heat absorbing glass
j：Obayashi Corporation

a：岐阜メモリアルセンター
b：岐阜県岐阜市
c：岐阜市
d：日建設計／高山順行
e：日建設計／葛原定次
f：大日本・土屋組・大武組 J.V.
g：SS名古屋
h：大小相似形のツインドーム型体育館と武道館を併設する総合スポーツセンター。周辺環境に配慮。
i：屋根：ステンレス鋼板フッ素樹脂塗装，外壁：磁器質タイル，アリーナ床：桜フローリング
j：日建設計
●

a : GIFU MEMORIAL CENTER
b : Gifu-shi, Gifu
c : Gifu City
d : Nikken Sekkei/Yoriyuki Takayama
e : Nikken Sekkei/Sadatsugu Kuzuhara
f : Dainippon, Tsuchiya-gumi & Otake-gumi, J.V.
g : SS Nagoya
h : An all-sports center accomodating similarly shaped large and small twin dome-style gymnasiums and martial art arenas. Great care was afforded to the surroundings.
i : Roof : stainless steel sheets with flouride resin coating, external walls : porcelain tiles, arena floors : cheery tree flooring
j : Nikken Sekkei Ltd.

a：名古屋国際会議場 白鳥センチュリープラザ
b：愛知県名古屋市
c：名古屋市
d：日建設計／高山順行
e：日建設計／葛原定次
f：大成・鴻池組・安藤・浅沼・徳倉 J.V.
g：SS 名古屋
h：国際会議，イベント，展示に利用される複合施設。
'89年世界デザイン博覧会終了後国際会議場に改称。
i：屋根：鉛ステンレス複合板，外壁：磁器質タイル，
建具：耐候性鋼フッ素樹脂塗装
j：日建設計
●

a: NAGOYA CONGRESS-SHIRATOEI
CENTURY PLAZA
b: Nagoya-shi, Aichi
c: Nagoya City
d: Nikken Sekkei/Yoriyuki Takayama
e: Nikken Sekkei/Sadatsugu Kuzuhara
f: Taisei, Konoike-gumi, Ando, Asanuma &
Tokura, J.V.
g: SS Nagoya
h: This is a multi-purpose complex facility
used for international congresses, events and
exhibitions. Its name was changed to the
International Congress Center since the closing
of the World Design Exhibition '89.
i: Roof : leaded stainless mixed sheet,
external walls : porcelain tiles, fixtures : all-
season flouride resin coating
j: Nikken Sekkei Ltd.

a：星和台ファミリーホール
b：兵庫県神戸市
c：新星和不動産
e：藤縄正俊，大橋眞由美
f：大林組
g：ナトリ光房
h：星和台地区のシンボルとなるよう三角屋根のトップ
ライトで特徴づけ，温かみのある建物とした。
i：吹き付けタイル
j：大林組本店一級建築士事務所

● a：SEIWADAI FAMILY HALL
b：Kobe-shi, Hyogo
c：Sinseiwa Fudosan Co.,Ltd.
e：Masatoshi Fujinawa, Mayumi Ohashi
f：Obayashi Corporation
g：K.Natori Studio
h：It is designed to be a warmhearted building
which is characterized with a roof light of a
peaked roof.
i：Sprayed tile
j：Obayashi Corporation

a：トヨタオート大阪サンダンス2000
b：大阪府守口市
c：トヨタオート大阪
e：竹中工務店設計部/狩野忠正，上坂脩
f：竹中工務店
g：大嶋勝寛
h：昼間は白く輝く外壁が大空の表情を映し，夜間は発光体となって車のシルエットを映しライティング・メッセージを伝える。
i：外壁：アルミカーテンウォール，ポリエステル繊維入り合わせガラス
j：竹中工務店
●
a：TOYOTA AUTO OSAKA SUNDANCE 2000
b：Moriguchi-shi, Osaka
c：Toyota Auto Osaka Corporation
e：Takenaka Corporation, Building Design Dep./Tadamasa Kano, Osamu Kosaka
f：Takenaka Corporation
g：Katsuhiro Oshima
h：The sparkling white exterior walls reflect the expressions of the sky during the day, and will become luminuos bodies during the night to reflect the silhoette of the cars delivering written messages.
i：Exterior walls : aluminium curtain wall, Polyeater fiber combined glass
j：Takenaka Corporation

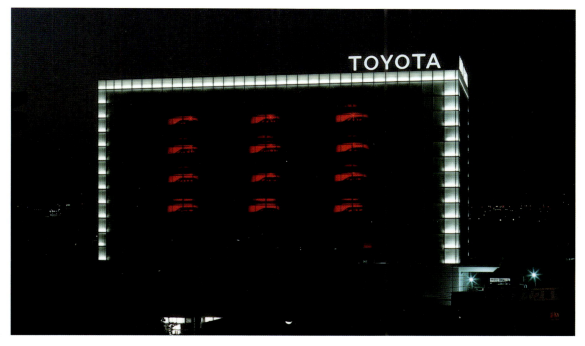

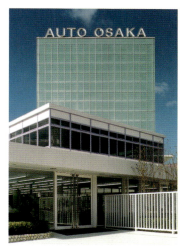

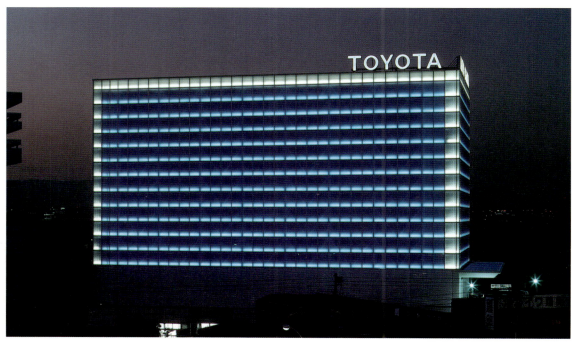

a：オムロン中央研究所3号館
b：京都府長岡京市
c：オムロン
e：竹中工務店設計部/野村充，角島健二
f：竹中工務店
g：ナトリ光房/名執一雄
h：芝生・テラス・水路による中庭と，コラージュによる統合を目指した住宅地に面するファサード。
i：タイル，アルミ，ハーフミラーガラス，枕木，御影石
j：竹中工務店
●

a : OMRON CENTRAL RESEARCH LABORATORY BUILDING NO.3
b : Nagaokakyo-shi, Kyoto
c : Omron Corporation
e : Takenaka Corporation, Building Design Dep./Michiru Nomura, Kenji Kadoshima
f : Takenaka Corporation
g : Natori Studios/Kazuo Natori
h : The facade facing the residential area was aimed at the creation of a courtyard by the addition of a Lawn, terrace and water conduit and integration by collage.
i : Tiles, aluminium, half-mirror glass, crossties, granite stone
j : Takenaka Corporation

a：西島プレス白井工場
b：千葉県印旛郡
c：西島プレス
e：竹中工務店設計部/神田孜，上野教人，坂口彰
f：竹中工務店
g：タミ・アート
h：屋根形態を緩やかなカーブの連続性で構成することにより，建物の独自性と環境との共存をさせた。
i：屋根：金属板瓦棒葺（ガルバニウム鋼板）/外壁：コンクリート打っ放し，撥水剤塗布
j：竹中工務店
●

a : NISHIJIMA PRESS SHIROI FACTORY
b : Inba-gun, Chiba
c : Nishijima Press
e : Takenaka Corporation, Building Design Dep./Tsutomu Kanda, Yoshito Ueno, Akira Sakaguchi
f : Takenaka Corporation
g : Tami Art
h : The uniqueness of this building and ist surroundings were coordinated by forming a gently curving continuity to the roof.
i : Roof : Metal sheet tile roofing (galvanized steel plate)/Exterior wall : Exposed concrete, spead waterproofing
j : Takenaka Corporation

a：サッポロビール千葉工場ゲストハウス
b：千葉県船橋市
c：サッポロビール
e：大成建設設計本部/東海林孝司，碇屋雅之，前田康貴
f：大成・鹿島・西松・三井建設共同企業体
g：ハットリスタジオ
h：全天候型のシアターを内包しており，21世紀へ向けて最新鋭工場のシンボル性と話題性を意図している。
i：チタン，フッ素樹脂焼付鋼板パネル，ハーフミラーガラス，焼過レンガ，御影石
j：大成建設設計本部
●

a: SAPPORO BEER CHIBA BREWERY, GUEST HOUSE
b: Funabashi-shi, Chiba
c: Sapporo Breweries Ltd.
e: Taisei Corporation Design and Proposal Division, Takashi Shoji, Masayuki Ikariya, Yasutaka Maeda
f: Taisei/Kajima/Nishimatsu/Mitsui, J.V.
g: Hattori Studio
h: A guest house including all weather theater is designed to be symbolized and to became the topic of a talk as a state-of-the-art factory toward 21 century.
i: Titanium, fluororesin baked steel plate panel, half mirror glass, clinker brick, granite
j: Taisei Corporation Design and Proposal Division

a：サッポロビール北海道工場
b：北海道恵庭市
c：サッポロビール
e：大成建設設計本部/塩津一興，手塚清次郎，宮原
義治/(造園担当)アーク・クルー
f：大成建設・伊藤組・西松建設・前田建設・鴻池組 J.V.
g：高崎清治(山田商会)
h：豊かな自然環境を創り出し，開かれた庭園工場を
建設。
i：断熱軽量長尺サンドイッチパネル/樹木：ハルニレ，
イチイ，カラマツ，ナナカマド
j：大成建設設計本部

●

a：SAPPORO BEER HOKKAIDO BREWERY
b：Eniwa-shi, Hokkaido
c：Sapporo Breweries Ltd.
e：Taisei Corporation Design and Proposal
Division, Kazuoki Siotu, Seijirou Tezuka,
Yoshiharu Miyahara/Ark Crew
f：Taisei, Ito, Nishimatsu, Maeda, Konoike J.V.
g：Seiji Takasaki (Yamada Syokai)
h：An opened garden brewery which is
familiarized by employees as well as local
which harmonizes with nature and creates
abundant natural environment.
i：Heat insulation lightweight long sandwich
panel/Arbores : Ulmus davidiana, Japanese
yew, Japanese larch, Mountain ash
j：Taisei Corporation Design and Proposal
Division

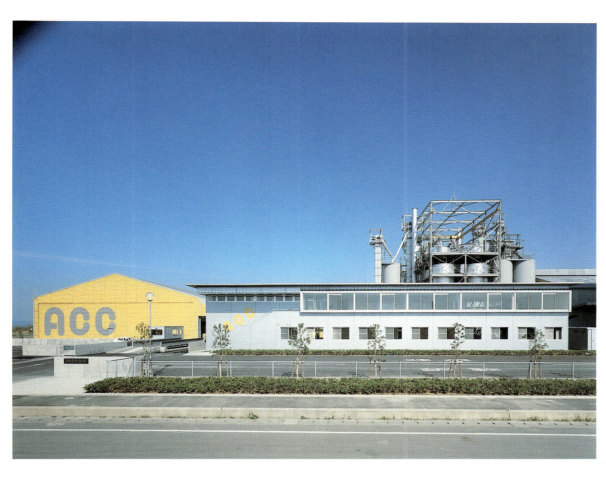

a：アライカーボン　八代工場
b：熊本県八代市
c：アライカーボン
e：野中建築事務所
f：和久田建築
g：伊藤直明
h：黒い粉塵で汚れ易いカーボン工場を材質とカラコンにより変化させ，臨海工業地の色彩環境モデルとして提案。
i：スチール，アルミ，スレート
j：野中建築事務所，野中卓

●

a：ARAI CARBON YATSUSHIRO PLANT
b：Yatsushiro-shi, Kumamoto
c：Arai Carbon Co.,Ltd.
e：Takashi Nonaka Architect
f：Wakuda Kensetsu Co.,Ltd.
g：Naoaki Ito
h：A carbon factory is changed from a place easy to be smudget with black dust into a color environment model of coastel industrial location.
i：Steel, aluminum, slate
j：Nonaka Architect & Associates, Takashi Nonaka

a：中外製薬宇都宮工場
b：栃木県宇都宮市
c：中外製薬
d：鹿島建設/羽生昌弘
e：鹿島建設/寺島振介，小野寺康夫
f：鹿島建設
g：SS東京
h：清原工業団地に建設された中外製薬の最新鋭主力工場。池を中心に21世紀を展望した環境計画内容。
i：コンクリートブロック，アスコン，メタセコイア，ヤマモミジ，他
j：鹿島建設
●
a：CHUGAI PHARMACEUTICAL COMPANY, UTSUNOMIYA WORKS
b：Utsunomiya-shi, Tochigi
c：Chugai Pharmaceutical Company
d：Kajima Corporation/Masahiro Hanyu
e：Kajima Corporation/Shisuke Terashima, Yasuo Onodera
f：Kajima Corporation
g：SS Tokyo
h：The latest flag-ship factory of the Chugai Pharmaceutical Company built in the Kiyohara Industrial Estate. The concept concentrates on the environmnetal plan for the 21st century which is based on the theme of ponds.
i：Concrete blocks, asphalt concrete, Metasekoia, mountain maple, etc.
j：Kajima Corporation

a：スタンレー電気秦野製作所新2号館
b：神奈川県秦野市
c：スタンレー電気
e：竹中工務店設計部/神田孜，上野教人，吉田昇
f：竹中工務店
g：吉村行雄
h：地域のランドマークとなり得る職場環境を備えた，技術革新に対応できるハイテク工場を目指した。
i：鉄骨フレーム，磁器タイル，アルミカーテンウォール，熱線反射ガラス
j：竹中工務店
●
a：STANLEY HADANO FACTORY NEW NO.2 BUILDING
b：Hadano-shi, Kanagawa
c：Stanley Electric Co.,Ltd.
e：Takenaka Corporation, Building Design Dep./Tsutomu Kanda, Yoshito Ueno, Noboru Yoshida
f：Takenaka Corporation
g：Yukio Yoshimura
h：The target here was focused on construction of a high tech factory which stood on the correct site to become an area landmark and which could cope with up and coming technical reforms.
i：Reinforcement frames, ceramic tiles, aluminium curtain wall, heat-reflecting glass
j：Takenaka Corporation

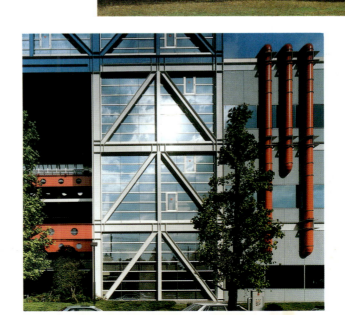

a：アシックススポーツ工学研究所人材開発センター
b：兵庫県神戸市
c：アシックス
e：竹中工務店設計部/柏木浩一，酒井利行
f：竹中工務店
g：村井修
h：建築とアートの融合をテーマに，スポーツと健康の
環境を創出した。
i：コンクリート打っ放し撥水剤塗布
j：竹中工務店
●
a：ASICS CORPORATION RESEACH
INSTITUTE OF SPORTS SCIENCE PERSNNEL
DEVELOPMENT CENTER
b：Kobe-shi, Hyogo
c：Asics Corporation
e：Takenaka Corporation, Building Design
Dep./Koichi Kashiwagi, Toshiyuki Sakai
f：Takenaka Corporation
g：Osamu Murai
h：With the integration of architecture and as a
theme, a sports and helth environment was created.
i：Xeposed concrete water proof spread
j：Takenaka Corporation

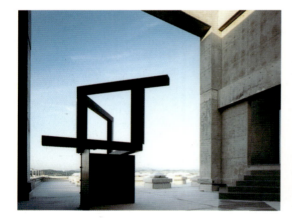

a:日立基礎研究所
b:埼玉県鳩山市
c:日立基礎研究所
d:日立建設
e:鹿島建設／上野卓二，小野寺康夫，羽生昌弘，日立建設，彫刻：児玉慎憲，児玉もえみ
f:鹿島建設
g:川澄建築写真事務所，SS 東京
h:埼玉県の景観賞受賞。周囲の環境を大切にしたランドスケープ計画。池は白鳥もおよいでいる。
i:コンクリートブロック，タイル，計画地の自然植生にそった選定樹種
j:鹿島建設

●

a: HITACHI ADVANCED RESEARCH LABORATORY
b: Hatoyama-shi, Saitama
c: Advanced Research Laboratory, Hitachi Ltd.
d: Hitachi Architects & Engineers Co.,Ltd.
e: Kajima Corporation/Takuji Ueno, Yasuo Onodera, Masahiro Hanyu, Hitachi Architects & Engineers Co.,Ltd., Sculptor : Shinken Kodama, Moemi Kodama
f: Kajima Corporation
g: Kawasumi Architectural Photograph Office, SS Tokyo
h: The best view of the Saitama Prefecture award-winning work. The landscaping was designed not to spoil the surrounding environment. Swans are kept in the pond.
i: Concrete blocks, tiles, a selection of plants were chosen to suit the surrounding nature
j: Kajima Corporation

a：日本合成ゴム筑波研究所
b：茨城県つくば市
c：日本合成ゴム
d：宇口進
e：菊池東紀男, 黒野雅好, 枡野俊明（日本造園設計）
f：鹿島建設：建築, 王子緑化：造園
g：川澄建築写真事務所
h：研究活動を活性化するためにアトリウムをもったインターラクションスペース（交流の場）と、これに空間的に連続する日本庭園のあるランドスケープ造りを目指した。
i：外壁：磁器質タイル貼り
j：日本設計
●
a：JAPAN SYNTHETIC RUBBER TUKUBA
RESEARCH CENTER
b：Tsukuba-shi, Ibaraki
c：Japan Synthetic Rubber Co.,Ltd.
d：Susumu Uguchi
e：Tokio Kikuchi, Masayoshi Kurono, Toshiaki
Masuno（Japan Landscape Consultants）
f：Kajima Corp.：Construction, Ohji Ryokka：
Landscape
g：Kawasumi Architectural Photograph Office
h：An interaction space having an atrium and
landscape having a Japanese garden which
spatially continue are made to activate
research activities.
i：Exterior wall：Vitreous tile
j：Nihon Sekkei, Inc.

a：大石田町クロスカルチャープラザ
b：山形県大石田町
c：大石田町/土屋市蔵
d：三橋一彦
e：三橋一彦，藤野和男
f：城北建設，大場工務店 J.V.
g：三橋一彦
h：舟運文化の街大石田をテーマにデザインが歴史を物語り共感を与える建物を意図。
i：コンクリート，鉄骨，ヒバ，ステンレス防水漆喰
j：三橋建築設計事務所
●

a : OISHIDA-MACHI CROSS CULTURE PLAZA
b : Oishida-machi, Yamagata
c : Oishida-Machi/Ichizo Tsuchiya
d : Kazuhico Mihashi
e : Kazuhiko Mihashi, Kazuo Fujino
f : Zyohoku, Oba J.V.
g : kazuhiko Mihashi
h : It is designed to express the town's history and sympathy with such a nice theme ohishda is a town having a culture of trading ships.
i : Concrete, structural steel, white-cedar leaves, stainless water-proof mortar
j : Mihashi Architectural Office

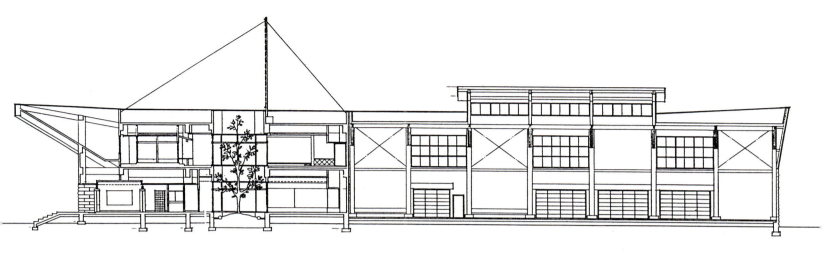

a：奥郷屋敷
b：香川県木田郡
c：奥郷屋敷
e：竹中工務店設計部/狩野忠正，砂川裕幸，野田隆史
f：竹中工務店
g：東出清彦
h：伝統的屋根材である本瓦葺きにより，中庭をもつ円形屋根という新しい表現を試みた。
i：本瓦，あじ石砂利，由良石，コンクリート
j：竹中工務店
●
a：OKU-GO YASHIKI
b：Kida-gun, Aichi
c：Oku-Go Yashiki
e：Takenaka Corporation, Building Design Dep./Tadamasa Kano, Hiroyuki Sunagawa, Takashi Noda
f：Takenaka Corporation
g：Kiyohiko Higashide
h：This house with a courtyard and new style of demed roof was created by the use of traditional Japanese roofing tiles.
i：With the integration of architecture and art as a theme, a sports and health environment was created
j：Takenaka Corporation

a：セコム HD センター名張
b：三重県名張市
c：セコム
e：日建設計／谷口望，住友林業緑化（植栽）
f：竹中工務店
g：松村芳治
h：研究所と保養所が併存する。外部環境，特に植栽
を充実させ，アプローチを長くとり入口を分離させた。
i：外壁：磁器質タイル，屋根：特殊セメント成型板，
舗装：骨材入りアスファルト・ブラスト仕上げ
j：日建設計
●

a : SECOM HD CENTER, NABARI
b : Nabari-shi, Mie
c : Secom Co.,Ltd.
e : Nikken Sekkei/Nozomu Taniguchi,
Sumitomo Ringyo Landscaping Co.,Ltd.
f : Takenaka Corporation
g : Yoshiharu Matsumura
h : This is a combination of a research institute
and a sanatorimu. The rich surroundings, with
emphasis on the plants, and the creation of the
long approach has produced tow seperate
entrances.
i : Enternal walls : stoneware (walling) tiles,
roof : special formation concrete boards,
paving : blast-finished aggregate contained
asphalt
j : Nikken Sekkei Ltd.

a：住友生命 OBP プラザビル いずみホール
b：大阪府大阪市
c：住友生命保険相互会社
d：日建設計／与謝野久
e：日建設計／中西博隆，花田佳明
f：竹中工務店・鹿島建設・熊谷組・住友建設 J.V.
g：黒田青巌
h：水・緑・光をモチーフとし，ビル内のいずみホールで
の演奏会の余韻を楽しめる外部空間を意図した。
i：床／100角磁器タイル，ボーダー150角磁器タイル，
噴水・滝まわり／花こう岩
j：日建設計
●
a：SUMITOMO OBP PLAZA BUILDING, IZUMI HALL
b：Osaka-shi, Osaka
c：Sumitomo Life Insurance Company
d：Nikken Sekkei/Hisashi Yosano
e：Nikken Sekkei/Hirotaka Nakanishi, Yoshiaki Hanada
f：Takenaka, Kajima, Kumagaya-gumi and
Sumitomo Kensetsu, J.V.
g：Seigan Kuroda
h：This outside space has been designed to
enjoy the lingering sounds of concerts coming
from within Izumi Hall amongst the motifs of
water, greenery and light.
i：Floor/100mm square ceramic tiles, border/
150mm square ceramic tiles, fountain near the
waterfall/granite veneer
j：Nikken Sekkei Ltd.

a：福井県自然保護センター
b：福井県大野市
c：福井県
d：林行正
e：千名良樹，小澤浩利，藤本聖徳，柳澤ゆき
f：日展
g：関西カラー写真
h：経ヶ岳のブナ林を，直径7m，高さ8mの円筒の中に
GLをH3mに設定し，そのまま表現した。
i：松/スチール/FRP（地層）
j：日展
●

a : NATURE CONSERVATION CENTER
b : Ono-shi Fukui
c : Fukui-ken
d : Yukimasa Hayashi
e : Yoshiki Senna, Hirotoshi Ozawa, Kiyonori
Fujimoto, Yuki Yanagisawa
f : Nitten Co.,Ltd.
g : Kansai Color Shasin Co.,Ltd.
h : The Japanese beech grove of Kyogatake is
put in a cylinder.It measure is dia.7m, h.8m.GL
is set H.3m.
i : Pine, steel, FRP (stratus)
j : Nitten Co.,Ltd.

a：ぐんまこどもの国児童会館
b：群馬県太田市
c：群馬県
d：山口広喜
e：藤森宣光，井下公彦
f：丹青社第2事業本部文化空間デザイン統括部
g：小原勉
h：科学の要素を織り込んだ遊びの施設のため，おもち
や箱的なデザイン・カラーリングを考慮した。
i：ステンレス，スチールメラ焼，ガラス，鏡面ステン
レス，FRP，フレキシブルボード
j：丹青社
●

a : GUNMA CHILDREN'S MUSEUM
b : Ohta-shi, Gunma
c : Gunma Prefecture
d : Hiroki Yamaguchi
e : Nobumitsu Fujimori, Kimihiko Inoshita
f : Tanseisha Co.,Ltd.
g : Tsutomu Ohara
h : Toy-box-like design and coloring are
contrived for a play facility weaving scientific
elements.
i : Stainless, Steel, glass, miror surfaced
stainless, FRP, flexible board
j : Tanseisha Co.,Ltd.

a：大栄教育システム名張研修所
b：三重県名張市
c：大栄教育システム
e：竹中工務店設計部/門川清行，清水弘之
f：竹中工務店
g：竹中工務店/吉村行雄，古川泰造
h：シンプルかつストレートな構成と表現を基調とし，のびのびとした空間を実現している。
i：コンクリート打っ放し，エマルジョンペイント，ナラフローリング（オイルフィニッシュ）
j：竹中工務店
●
a：DAIEI NABARI EMINAR HOUSE
b：Nabari-shi, Mie
c：Daiei Education System Co.,Ltd.
e：Takenaka Corporation, Building Design Dep./Kiyoyuki Kadokawa, Hiroyuki Shimizu
f：Takenaka Corporation
g：Takenaka Corporation/Yukio Yoshimura, Taizo Furukawa
h：This simple and straight basic construction has manged to create a free a relaxed space.
i：Exposed concerte, emulsion paint, Japanese oak (oil finished)
j：Takenaka Corporation

a：富士ゼロックス，徳島保養所
b：徳島県徳島市
c：富士ゼロックス
e：大成建設設計本部/合川彰夫，川村光明
g：富士ゼロックス
h：屋根づたいの変形敷地と鳥獣保護区など厳しい条件のため，平面計画に頭をいためた。
i：屋根：銅板/外壁：庵治石積み，一部スタッコ/内壁：庵治石積
j：大成建設設計本部
●

a：FUJI XEROX TOKUSHIMA SANATORIUM
b：Tokushima-shi, Tokushima
c：Fuji Xerox Corp.
e：Taisei Corporation Design and Proposal Division, Akio Aikawa, Mitsuaki Kawamura
g：Fuji Xerox Corp.
h：A plane plan was the most difficult plan because of the strict conditions such as deformation site which goes along with a spine and it is in a wild life sanctuary.
i：Roof : Copper sheets/Exterior wall : Stone masonry partly stucco/Interior wall : Stone masonry
j：Taisei Corporation Design and Proposal Division

a：積水ハウス総合住宅研究所
b：京都府相楽郡
c：積水ハウス
d：日建設計／小角亨
e：日建設計／猪子順、藤井昭造，造園コンサルタント／中村一，空充秋，壁面カラーデザイン／中川千早
f：竹中工務店・前田建設工業 J.V.
g：東出清彦
h：住宅地に建つ郊外型の研究所。異業種13社の研究所団地（ハイタッチ・リサーチパーク）も中核的施設。
i：外壁／磁器質タイル（ラスター），吹抜け空間壁／モルタル EP-II
j：日建設計

●

a：SEKISUI HOUSE-COMPERHENSIVE HOUSING R & D INSTITUTE
b：Soraku-gun, Kyoto
c：Sekisui House, Ltd.
d：Nikken Sekkei/Toru Kosumi
e：Nikken Sekkei/Jun Inoko, Shozo Fujii, Garden Design Consultant/Hajime Nakamura, Mitsuaki Sora, Chihaya Nakagawa
f：Takenaka Corporation & Maeda Construction J.V.
g：Kiyohiko Higashide
h：A suburban research institute built in a residential area.A core facility of the 13 differnt companies that are housed within this reaearch and housing eastate (high-touch resarch park).
i：Eterior walls/ceramic-style tiles (lustre), viod space wall/mortal EP-II
j：Nikken Sekkei Ltd.

a：十六銀行 研修所
b：岐阜県岐阜市
c：十六銀行
d：日建設計／南石周作
e：日建設計／丸山茂，石倉義行
f：清水建設
g：SS 名古屋
h：この研修所は，企業研修と相互コミュニケーションの施設で，のびやかさと清楚さがテーマ。
i：床：タイルカーペット，壁：多彩模様吹き付け，天井：岩綿吸音板ペンキ仕上げ
j：日建設計

●

a：JUROKU BANK EDUCATION CENTER
b：Gifu-shi, GIfu
c：The Juroku Bank Co.,Ltd.
d：Nikken Sekkei/Shusaku Nanseki
e：Nikken Sekkei/Shigeru Maruyama, Yoshiyuki Ishikura
f：Shimizu Corporation
g：SS Nagoya
h：This research institute has been built as an in-service training institute for companies and as a mutual communications facility under the theme of comfortableness and cleanliness.
i：Floor：tile carpeting, walls：multi-colored patterned spray, ceiling：rock-wool sound absorbtion sheet with painted finish
j：Nikken Sekkei Ltd.

a：日新製鋼呉製鉄所若葉研修センター
b：広島県呉市
c：日新製鋼
e：竹中工務店設計部/井上軍，小林泰樹
f：竹中工務店
g：国際写真
h：これからの研修施設像として，リゾート性の展開を
テーマとした。
i：外壁：ステンレス，45角磁器モザイクタイル/サッ
シ：ステンレス，アルミ
j：竹中工務店
●
a：NISSHIN STEEL CO.,LTD., KURE WORKS,
WAKABA TRAINING CENTER
b：Kure-shi, Hiroshima
c：Nisshin Steel Co.,Ltd.
e：Takenaka Corporation, Building Design
Dep./Isao Inoue, Yasuki Kobayashi
f：Takenaka Corporation
g：Kokusai Shashin
h：The future image of this training facility was
planned around a resort-style development.
i：Exerior walls : stainless steel, 45 square
ceramic mosaic tile/Sashes : stainless
aluminium
j：Takenaka Corporation

a：玉木病院
b：山口県萩市
c：玉木病院
e：竹中工務店設計部/佐藤正二，三木重典
f：竹中工務店
g：国際写真
h：中庭を中心に展開したインテリアは，自然（光，風，緑）と融合し病院の既成イメージを一新した。
i：外壁：45角磁器モザイクタイル/中庭床：100角磁器タイル
j：竹中工務店
●
a：TAMAKI HOSPITAL
b：Hagi-shi, Yamaguchi
c：Tamaki Hospital
e：Takenaka Corporation, Building Design Dep./Shoji Sato, Shigenori Miki
f：Takenaka Corporation
g：Kokusai Shashin
h：This interior was coordinated to integrate with the nature of the courtyard (light, wind, greenery), changing the image of the ready-made hospital.
i：Exterior wall : 45 sq.ceramic mosaic tilles/ Central flooring : 100 sq.ceramic tilles
j：Takenaka Corporation

a：カネディアン アカデミィ
b：兵庫県神戸市
c：カネディアン アカデミィ
e：竹中工務店設計部/小川清一，前川清，後藤浩美
f：竹中工務店
g：村井修
h：建物中心二階のプラザは，平坦な埋立地において丘をイメージし，陽光と風が通る明るい学校とした。
i：外壁：タイル，アルミ/プラザ：タイル，吹き付けタイル，アルミ，ステンレス
j：竹中工務店
●

a：CANADIAN ACADEMY
b：Kobe-shi, Hyogo
c：Canadian Academy
e：Takenaka Corporation, Building Design Dep./Seiichi Ogawa, Kiyoshi Maekawa, Hiromi Goto
f：Takenaka Corporation
g：Osamu Murai
h：The second story plaza in the middle of the building was imagined as a hill on plain reclaimed land in order to create a bright school with plenty of sun and ventilation.
i：Exterior walls : Tiles, aluminium/Plaza : tiles, spray tiles, aluminium, stainless steel
j：Takenaka Corporation

a：代々木ゼミナール岡山校
b：岡山県岡山市
c：代々木ゼミナール
e：大成建設設計本部/井上隆夫，長嶋達也
f：大成建設
g：大宣
h：エントランス部にポケットパークを設け，町並みの景観にヒューマニティスペースを主張した。
i：花こう石，磁器タイル打込PC板，アルミサッシ，ステンレス
j：大成建設設計本部
●
a：YOYOGI SEMINAR OKAYAMA
b：Okayama-shi, Okayma
c：Yoyogi Seminar
e：Taisei Corporation Design and Proposal Division, Takao Inoue, Tatsuye Nagashima
f：Taisei Corporation
g：Daisen
h：A pocket park is made at the entrance of the building to represent a humanity space in the scene of the streets.
i：Granite, PC panel with vitreous tile, aluminum sash, stainless
j：Taisei Corporation Design and Proposal Division

a：日之影町立日之影中学校
b：宮崎県日之影町
c：日之影町
d：小嶋凌衛
e：小嶋凌衛
f：五洋・木田建設共同企業体，上田・田村建設共同企業体
g：小嶋凌衛
h：山深い自然のなかに欧風の城と雁行平面による平面の変化により，地域の拠所とインパクトを求めた。
i：陶器瓦/フッ素樹脂鋼板/アルミサッシ/コンクリート打っ放し/内装：杉，檜
j：小嶋凌衛建築設計事務所，小嶋凌衛
●
a：HINOKAGE JUNIOR HIGH SCHOOL
b：Hinokage-cho, Miyazaki
c：Hinokage-cho
d：Ryoei Kojima
e：Ryoei Kojima
f：Goyo/Kida J.V., Ueda/Tamura J.V.
g：Ryoe Kojima
h：It is designed to give an impact and become a landmark in the area by the change of plane caused between an European castle in the heart of a mountain and stagger plane.
i：Chinaware tile, fluoroplastic steel plate, aluminum sash, exposed concrete surface. Interior : Cedar, cypress
j：Ryoe Kojima and Associates

a：北海道安達学園
b：北海道札幌市
c：学校法人東京学園
d：中本嘉彦
e：吉村祐子
f：大成建設
g：やまだ商会
h：札幌大通り公園の前の専修学校。前庭の屋外空間
と吹抜け内部空間とのバランスがおもしろい。
i：外部：せっ器質ハツリタイル貼り/ロビー，ホール：
御影石水磨仕上げパターン貼り
j：エー・アンド・エー
●
a：HOKKAIDO ADACHI GAKUEN
b：Sapporo-shi, Hokkaido
c：School Corporation Tokyo Gakuen
d：Yoshihiko Nakamoto
e：Yuko Yoshimura
f：Taisei Kensetu Corporation
g：Yamada Corporation
h：A special school in front of the Sapporo
Odori park.The balance between an entrance
court as out space and an atrium as interior
space seems to be interesting.
i：Exterior : stone ware chipping tile/Lobby &
hall : Granite rubbing finish
j：A & A Co.,Ltd.

a：めぐみ学園リフレッシュ工事
b：京都府長岡京市
c：学校法人めぐみ学園
d：蔭山利夫
e：真田元晴，由比清隆，有居徹彦，元井正太郎
f：乃村工藝社
g：村瀬武男
h：園児たちが日々の生活のなかで，たのしく，健康的
で明るい環境で学べ，快い印象を後々まで感じる空間を意図した。
i：外装/ボンタイル吹き付け加工，内装/EP塗装天
井，壁・床：フローリング，染木加工
j：乃村工藝社
●
a：MEGUMI KINDERGARTEN REFRESH PLAN
b：Nagaokakyo-shi, Kyoto
c：Megumi Kindergarten
d：Toshio Kageyama
e：Motoharu Sanada, Kiyotaka Uhi, Tetsuhiko
Arii, Shotaro Motoi
f：Nomura Co.,Ltd.
g：Takeo Murase
h：This space was designed for children to
spend time in learning happily in a heathly and
bright environment while leaving fond
memories behind.
i：Exterior decorations/Bon tiles with sprayed
finish, Interior decorations/EP painted ceiling
and walls, flooring with woody finish for the floors
j：Nomura Co.,Ltd.

a：東京写真専門学校四番町校舎
b：東京都千代田区
c：学校法人東京写真学園
d：中本嘉彦
e：梨田俊次
f：住友建設
g：SSグループ
h：密集した都心住居群の中の専修学校。ハードな機
材とソフトなフォルム，そして緑の庭が特長。
i：外壁：50角磁器質タイル貼り，一部ガラスブロック
j：エー・アンド・エー
●
a：TOKYO PHOTO SPECIAL SCHOOL
b：Chiyoda-ku, Tokyo
c：Tokyo Shashin Gakuen Corporation
d：Yoshihiko Nakamoto
e：Shunji Nashida
f：Sumitomo Kensetsu Corporation
g：SS Group
h：A special school in a dense residential
district of the center of the Metropolis.Features
are hard materials, a soft form and green rich garden.
i：Exterior wall : Vitreous tile 50 mm square
(partly glass block)
j：A & A Co.,Ltd.

a：豊橋市自然史博物館
b：愛知県豊橋市
c：豊橋市
d：日建設計/川瀬友造
e：日建設計/日比野万喜男
f：鴻池組・小野 J.V.
g：SS 名古屋
h：この建物は、実大骨格標本アナトザウルスを中心に古生代・中生代・新生代を時系列で展示している。
i：外壁：特殊樹脂型枠コンクリート打っ放し、床：ビニールシート、壁・天井：ペンキ仕上げ
j：日建設計
●
a：TOYOHASHI MUSEUM OF NATURAL HISTORY
b：Toyohashi-shi, Aichi
c：Toyohasi City
d：Nikken Sekkei/Tomozo Kawase
e：Nikken Sekkei/Makio Hibino
f：Konoike-gumi & Ono, J.V.
g：SS Nagoya
h：With a life-size skeletal specimen of the Anatosaurus as a centerpiece, the Paleozoic, Mesazoic and Cenzoic ages are displayed in a time series within this building.
i：External walls : special resin framed exposed concrete, floor : vynal sheeting, walls and ceiling : paint finish
j：Nikken Sekkei Ltd.

a：花王「清潔と生活」小博物館
b：東京都墨田区
c：花王，凸版印刷
d：大森文樹
e：小澤浩利，柳澤ゆき，吉澤久枝
f：日展
g：首藤将夫
h：「清潔と生活」のテーマにそって，「清潔な空間」造りを試みた。
i：スモールパンチングメタル，大理石
j：日展
●
a：KAO CLEANLINESS AND LIFE EXHIBITION
b：Sumida-ku, Tokyo
c：Kao Corporation, Toppan Printing Co.,Ltd.
d：Fimiki Ohmori
e：Hirotoshi Ozawa, Yuki Yanagisawa, Hisae Yoshizawa
f：Nitten Co.,Ltd.
g：Masao Shudo
h：It is designed to make a clean space along with the theme Cleanliness and life.
i：Small punching metal, marble
j：Nitten Co.,Ltd.

a：トヨタ博物館
b：愛知県長久手町
c：トヨタ自動車
d：日建設計/南石周作
e：トヨタ自動車/田中勝男
f：竹中工務店
g：センターフォート
h：長方形のサーキットを思わせる平面形で中央に全階を貫くアトリウムをもつ本格的自動車博物館。
i：床：カーペットタイル（ラスター），壁・天井：多彩模様吹き付け
j：日建設計
●
a：TOYOTA AUTOMOBILE MUSEUM
b：Nagakute-cho, Aichi
c：Toyota Motor Corporation
d：Nikken Sekkei/Shusaku Nanseki
e：Toyota Motor/Katsuo Tanaka
f：Takenaka Corporation
g：Center Photo
h：A full-scale automobile museum in the shape of a plane, which reminds one of a rectangle circuit, and an atrium core throughout each floor.
i：Floor：carpet tiles, walld and ceiling：multi-colored patterned spray
j：Nikken Sekkei Ltd.

a：高知市自由民権記念館
b：高知県高知市
c：高知市
d：寺本敏則
e：日建設計／大野一男，八幡健志
f：鴻池組・新進建設 J.V.
g：黒田青巌
h：前面道路から敷地奥までの軸線の中央に吹抜けのアトリウムをとり，奥行のある構成とした。
i：外壁：石灰石，GRC磁器タイル打込み，GRC フッ素塗装，天然スレート
j：日建設計

●

a : KOCHI MUNICIPAL DEMOCRATIC
MOVEMENT MEMORIAL MUSEUM
b : Kochi-shi, Kochi
c : Kochi City
d : Toshinori Teramoto
e : Nikken Sekkei/Kazuo Ono, Kenji Yawata
f : Konoike-gumi & Shinshin Kensetsu, J.V.
g : Seigan Kuroda
h : A void atrium has been created in the
center of axis between the front road and the
back of the site in order to make a
construction with depth.
i : External walls : limestone, GRC porceline
tile coating, GRC flourine paint, natural slate
j : Nikken Sekkei Ltd.

a：福岡市博物館
b：福岡県福岡市
c：福岡市
d：中沢文三
e：藤井明文，岩本勝也，森山純治
f：丹青社第2事業本部文化空間デザイン統括部
g：小原勉
h：2,200平方メートルでワンルームを「開けた明るい展示室」にするため，素材，形態，色彩，照明などの演出計画を行った。
i：高透過度ガラス，トープラストーン，スチール下地ベルビアン貼り
j：丹青社
●

a : FUKUOKA CITY MUSEUM
b : Fukuoka-shi, Fukuoka
c : Fukuoka City
d : Bunzo Nakazawa
e : Akifumi Fujii, Katsuya Iwamoto, Junji Moriyama
f : Tanseisha Co.,Ltd.
g : Tsutomu Ohara
h : Materials, a form, coloring, lighting, are designed in order to a 2,200m sq sized room to be an open and bright exhibition room.
i : High transmission glass, steal setting
j : Tanseisha Co.,Ltd.

a：北とぴあ科学館
b：東京都北区
c：東京都北区
d：山口勝
e：木戸康裕，甲田敦子，柳澤ゆき
f：日展
g：首藤将夫
h：質量をテーマとした体験，体感型の参加性を重視した，アミューズメント性のある学習博物館である。
i：スチール/ステンレス/ハーフミラー/ホストコンピュータ，端末コンピュータ
j：日展
●

a : HOKUTOPIA SCIENCE MUSEUM
b : Kita-ku, Tokyo
c : Kita-ku, Tokyo
d : Masaru Yamaguchi
e : Yasuhiro Kido, Atsuko Kouda, Yuki Yanagisawa
f : Nitten Co.,Ltd.
g : Masao Shudo
h : It is an amusable educational museum which respect for having experience of studying gravitation and participation in bodily sensation.
i : Steel, Stainless steel, half mirror, host computer and terminal computer
j : Nitten Co.,Ltd.

a：気の博物館
b：栃木県芳賀郡
c：共同宣伝
d：本松浩司
e：指田茂，川畑英幹
f：日展
g：首藤将夫
h：神秘的な「気」の力を映像・造形などの演出効果を駆使し，その存在をさまざまな角度から実証しようと試みた。
i：ガラス/スチール/石/FRP/タモ
j：日展
●
a：KI MUSEUM
b：Haga-gun, Tochigi
c：Kyodosenden Advertising Agency
d：Kouji Motomatsu
e：Shigeru Sashida, Hideki Kawabata
f：Nitten Co.,Ltd.
g：Masao Shudo
h：It is designed to demonstrate the existence of the mysterious KI power commanding stage effects such as reflections and modeling.
i：Glass, steel, stone, FRP, shelf
j：Nitten Co.,Ltd.

a：我孫子市 鳥の博物館
b：千葉県我孫子市
c：我孫子市教育委員会
d：林行正
e：池邊哲治，桐山昇一，山口美弘，片岡麻子
f：日展
g：首藤将夫
h：鳥についての歴史や生態学を踏まえながら，市の環境保護と人と鳥との共存の願いを唱う博物館。
i：建築：鉄筋コンクリート/内装：スチール等
j：日展
●

a：ABIKO CITY MUSEUM OF BIRDS
b：Abiko-shi, Chiba
c：Members of a board of education in Abiko City
d：Yukimasa Hayashi
e：Tetsuji Ikebe, Shoichi Kiriyama, Yoshihiro Yamaguchi, Asako Kataoka
f：Nitten Co.,Ltd.
g：Masao Shudo
h：A museum advocating a hope of the protection of environment and being coexist human being and birds taking into account the history and ecology of the birds.
i：Building : Reinforced concrete.Interior : Steel etc.
j：Nitten Co.,Ltd.

a：山形生涯教育センター
b：山形県山形市
c：山形県
d：遠藤英雄
e：中辻伸，石井泰二，金子清二，森川和法，（エンジニア）鈴木俊幸，向隆宏
f：乃村工藝社
g：大東正巳
h：ネオクラシックな表現を基調に，遊びの心象風景を描き出し，生涯教育の象徴とすることを意図した。
i：スチール，ガラス
j：乃村工藝社
●

a：YAMAGATA SYOGAI-KYOIKU CENTER
b：Yamagata-shi, Yamagata
c：Yamagata Pref.
d：Hideo Endo
e：Shin Nakatsuji, Taiji Ishii, Seiji Kaneko, Kazunori Morikawa, (Engineer) Toshiyuki Suzuki, Takahiro Mukai
f：Nomura Co.,Ltd.
g：Masami Daito
h：An image view of play has been drawn by using neo-classic expressionalism in order to make it a symbol of life-time education.
i：Steel, glass
j：Nomura Co.,Ltd.

a：神奈川県立金沢文庫
b：神奈川県横浜市
c：神奈川県
d：堀田勝之，後山悟
e：（プランナー）三輪祐児，西村敏信，武田博宣，伴泉，（プロモーター）松浦恒久，高橋昭伸
f：乃村工藝社
g：大東正巳
h：鎌倉時代の木の文化をデザインコンセプトとし，展示室，移動・収蔵方式で柱などを構成した。
i：ヒノキ，松，ツガ，シンチュウ，スチール，クロス，無反射ガラス
j：乃村工藝社
●

a：KANAGAWA PREFECTURAL KANAZAWABUNKO MUSEUM
b：Yokohama-shi, Kanagawa
c：Kanagawa Pref.
d：Katsuyuki Hotta, Satoru Ushiroyama
e：(Planner) Yuji Miwa, Toshinobu Nishimura, Hironobu Takeda, Izumi Ban, (Promoter) Tsunehisa Matsuura, Akinobu Takahashi
f：Nomura Co.,Ltd.
g：Masami Daito
h：In the display room, mobile and built-in methods were used to construct columns with cultural wood from the Kamakura Period as a design concept.
i：Hinoki, pine, Japanese hemlock, brass, steel, cloth, non-reflective glass
j：Nomura Co.,Ltd.

a：ウッドピアいわき
b：福島県いわき市
c：福島県いわき市
d：高野裕之
e：執行昭彦，（プランナー）相原慎一
f：乃村工藝社
g：大東正巳
h：いわき産「目兼杉」材をシステマティックに組立てた造作による体験型装置群で環境を構築した。
i：目兼杉，スチール，ポリカーボネート，エンボスゴムタイル
j：乃村工藝社
●

a：WOODPIA IWAKI
b：Iwaki-shi, Fukusima
c：Iwaki-City, Fukusima
d：Hiroyuki Takano
e：Akihiko Shigyo, (Planner) Shinichi Aihara
f：Nomura Co.,Ltd.
g：Masami Daito
h：Iwaki-produced Megane cedar was systematically assembled to produce devices rich in experience in order to construct this environment.
i：Megane ceder, steel, polycarbonate, embossed rubber
j：Nomura Co.,Ltd.

a：高岡市万葉歴史館
b：富山県高岡市
c：高岡市
d：乃村工藝社／（総合ディレクター）堀田勝之，宮武博彦，（コンセプトプランナー）大野昭夫，高橋成忠，井上富，望月昭，（アドバイザー）石原堅次，（展示設計ディレクター）津田雅人，原田豊，（展示設計プランナー）高橋信裕，相原慎一，（建築設計・監修）清家清，（建築設計ディレクター）角永博
e：（展示設計デザイナー）三好恵介，宮野哲也，（建築設計デザイナー）大萱喜知郎，（造園設計デザイナー）中島猛夫，（絵画）平山郁夫
f：乃村工藝社
g：大東正巳
h：展示・建築・造園各設計において，四季の万葉歌を主題にし，環境融合を計ることを設計意図とした。
i：スチール加工，塩地染色，四季の歌模型造形，遣唐使船模型，映像可動家持劇場
j：乃村工藝社

●

a：TAKAOKA CITY MANYO HISTOIRY MUSEUM
b：Takaoka-shi, Toyama
c：Takaoka City
d：Nomura Co.,Ltd./(Total Director) Katsuyuki Hotta, Hirohiko Miyatake, (Concept Planner) Akio Ono, Shigetada Takahashi, Yutaka Inoue, Akira Mochizuki, (Adviser) Kenji Ishihara, (Disply Design Director) Masato Tsuda, Yutaka Harada, (Disply Design Planner) Nobuhiro Takahashi, Shinichi Aihara, (Architectural Design & Supervision) Kiyoshi Seike, (Architectural Design director) Hiroshi Suminaga
e：(Disply Designer) Keisuke Miyoshi, Tetsuya Miyano, (Architectural Designer) Kichiro Ogaya, (Garden Designer) Takeo Nakajima, (Picture) Ikuo Hirayama
f：Nomura Co.,Ltd.
g：Masami Daito
h：In the process of each design element of the exhibition, architecture and landscaping, the Four Seasons sung Manyoshu was made into a theme to add to this environmental intergration.
i：Steel processing, dye, Four Seasons molding, Chinese ship model, the mobile reflection of the Iemochi Theater
j：Nomura Co.,Ltd.

a：屋久杉自然館
b：鹿児島県熊毛郡
c：屋久町
d：高野裕之
e：執行昭彦，（プランナー）相原慎一
f：乃村工藝社
g：大東正巳
h：展示から建築を発想し，展示・建築とも設計することで空間の求心性と遠心性を結合させた。
i：屋久杉，銘木単板，スチール，RC 打っ放し
j：乃村工藝社

●

a：YAKUSUGI MUSEUM
b：Kumage-gun, Kagosima
c：Yaku-machi
d：Hiroyuki Takano
e：Akihiko Shigyo, (Planner) Shinichi Aihara
f：Nomura Co.,Ltd.
g：Masami Daito
h：The architectural concept was taken from the display.The centripetal and centrifugal seperation of the space was combined by designing both the displays and building.
i：Yaku cedar, choice wood, steel, exposed RC
j：Nomura Co.,Ltd.

a：大阪市立科学館
b：大阪府大阪市
c：大阪市
e：環境開発研究所
f：竹中工務店・大林組・鴻池組・銭高組共同企業体
g：大島勝寛
h：敷地16,000平方メートルの80％を空地として計画
されており，都心の広場としての機能を意識している。
i：磁器タイル，御影石，インターロッキングブロック
j：竹中工務店
●
a：THE SCIENCE MUSEUM, OSAKA
b：Osaka-shi, Osaka
c：Osaka City
e：Environment Development Research Inc.
f：Takenaka Corporation, Ohbayashi-Gumi,
Kohnoike-Gumi, Zenidaka-Gumi J.V.
g：Katsuhiro Oshima
h：80 % of the site, measuring 16,000m2, has
been specially designed as open space to
serve as a midtown square.
i：Ceramic tiles, granite stone, interlocking rock
j：Takenaka Corporation

a：武石村ともしび博物館
b：長野県小県郡
c：武石村
d：藤井明文
e：岩本勝也，福田直人
f：須長毅
g：小原勉，スタジオ・ベン
h：壁面のスリガラスに幻想的な切り絵を走馬燈のよう
に浮かび上らせ，展示のシンボルとした。
j：丹青社
●
a：TAKESHI-MURA TOMOSHIBI MUSEUN
b：Ogata-gun, Nagano
c：Takeshi-mura
d：Akifumi Fujii
e：Katsuya Iwamoto, Naoto Fukuda
f：Tsuyoshi Sunaga
g：Tsutomu Ohara(Studio Ben Co.,Ltd.)
h：A visionary Japanese craft art floated on
frosted glass on the wall as if it was a
revolving lantern is symbolized of exhibition.
j：Tanseisha Co.,Ltd.

a：国立科学博物館サイエンス・シアター
b：東京都台東区
c：国立科学博物館
d：(企画・監修)国立科学博物館，森山晶夫，武居清，(テクニカル・ディレクター)中山隆
e：(テクニカル・デザイナー)梅沢誠一郎，刈屋珠明
f：乃村工藝社
g：大東正巳
h：雑木林の生態から6テーマを選び，造形，メカ，映像，音響，照明を演出する「可動演出装置」。テーマ/モグラ，カッコー，ゲンジボタル，オニグモ，カブトムシ，オオカマキリ。
j：乃村工藝社

● NATIONAL SCIENCE MUSEUM SCIENCE THEATER

a：NATIONAL SCIENCE MUSEUM SCIENCE THEATER
b：Taito-ku, Tokyo

c：National Sience Museum
d：(Planning & Supervision) National Sience Museum, Akio Moriyama, Kiyoshi Takei, (Technical Director) Takashi Nakayama
e：(Technical Designer) Seiichiro Umezawa, Shumei Kariya
f：Nomura Co.,Ltd.
g：Masami Daito
h：Six different modes of life form were chosen from the thicket of assorrted trees, and mobile devices such as modeling, mechanism, reflection, sound and lighting were used to control the production.Theme/moles, cuckoos, Genji-fire-flies, garden spiders, beetles, large preying mantis
j：Nomura Co.,Ltd.

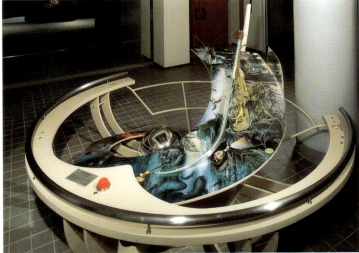

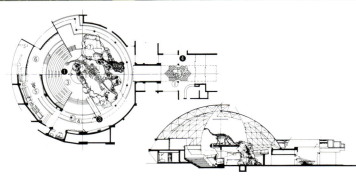

a：笠岡市立カブトガニ博物館
b：岡山県笠岡市
c：笠岡市
d：松村数正
e：野沢克昌，福島徹，栗原礼子
f：丹青社
g：小原勉，スタジオ・ベン
h：カブトガニを模した外観と，ドーム空間には「太古の世界」のイメージで展示・演出した。
j：丹青社

●

a：KASAOKA CITY HORSESHOE CRAB MUSEUM
b：Kasaoka-shi, Okayama
c：Kasaoka-City
d：Kazumasa Matsmura
e：Katsumasa Nozawa, Toru Fukushima, Reiko Kurihara
f：Tanseisha Co.,Ltd.
g：Tsutomu Ohara(Studio Ben Co.,Ltd.)
h：The exterior imitating a helmet crab and a dome space imaged ANCIENT WORLD are exhibited and produced.
j：Tanseisha Co.,Ltd.

a：芝浦スクエア
b：東京都港区
c：東電不動産管理，第一生命保険，日本生命，東信地所，中一地所，泉商事，ヨコオ
e：竹中工務店設計部/絹川正，木島幸一
f：竹中工務店
g：斎部功
h：インテリジェントビルと高層住宅の間に広場を介在させ，都心における職と住の共生のあり方を追求。
i：熱線反射ガラス，アルミ，ステンレス，石，磁器質タイル，コンクリート
j：竹中工務店
●

a : SHIBAURASQUARE
b : Minato-ku, Tokyo
c : The Tokyo Electric Power Real Estate Mamtenance Co., The Dai-ichi Mutual Life Insurance Co., Nipponu Life Insurance Co., Toshin Jisho Co.,Ltd., Nakaichi Real Estate, Izumi Shoji Co.,Ltd., Yoko'o Co.,Ltd.
e : Takenaka Corporation, Building Design Dep./Tadashi Kinukawa, Koichi Kijima
f : Takenaka Corporation

g : Isao Imbe
h : By laying out a plaza between theintelligent building and the multi-storied housing, a symbol indicating both work and living has been established.
i : Heat reflection glass, aluminium, stainless steel, stone, ceramic tiles, concrete
j : Takenaka Corporation

a：乃村工藝社大阪事業所ファサード＆ホール
b：大阪府大阪市
c：乃村工藝社
d：蔭山利夫、郷力憲治
e：乃村一級建築士事務所、建築委員会
f：乃村工藝社、建築メディア事業部、明和工務店
g：大東正巳
h：単純、明快、かつ意外性とユーモアをまじえたフォルム。情緒的でしなやかなフェイスとランドマーク効果の演出。
i：レフライトとカーテンウォール（アルミ）、両翼はステンレス、フッ素樹脂加工、床／大理石
j：乃村工藝社

a : NOMURA CO.,LTD., OSAKA OFFICE
b : Osaka-shi, Osaka
c : Nomura Co.,Ltd.
d : Toshio Kageyama, Kenji Goriki
e : Nomura Architecture Office, Architecture Committee
f : Nomura Co.,Ltd., Architecture Department, Meiwa Co.,Ltd.
g : Masami Daito
h : A forum that has mixed designs to summon up feeling of simpleness, clearness, the unexpected and humour.A production of passion and elegance with a landmark effect.
i : Lighting and curtain wall（aluminum）, stainless steel has been used for both wings. Flourine resin treament, marble stone for the floor
j : Nomura Co.,Ltd.

a：21世紀の安全をめざして
b：東京都千代田区
c：JR東日本
d：真鍋実
e：藤森宣光、山森博之
f：丹青社
g：小原勉
h：東京駅丸の内ドーム空間との融合を図り、パイプ組でスケール感を強調し、テーマの安全を色で表現した。
i：スチールパイプ／スチールパンチングメタル
j：丹青社

a : SAFETY CHALLENGE
b : Chiyada-ku, Tokyo
c : East Japan Railway Company
d : Minoru Manabe
e : Nobumitsu Fujimori, Hiroyuki Yamamori
f : Tanseisha Co.,Ltd.
g : Tsutomu Ohara
h : It is designed to be in harmony with Tokyo station Marunouchi doom space, to emphasize a big scale feeling by pipe construction and to express its them SAFETY using color.
i : Steel pipe, steel punching metal
j : Tanseisha Co.,Ltd.

a：福岡市博物館「歴史と水辺のふれあいの杜」
b：福岡県福岡市
c：福岡市
d：福岡市、佐藤総合計画
e：佐藤総合計画、イワオ・タイルデザインチーム
f：清水・大祥建設 J.V.
h：床面はざっくりとした素材感で暖か味を表現。天然原料を高温焼成しているため、耐久性は抜群。
i：透水性多孔質磁器
j：岩尾磁器工業

a : FUKUOKA-SHI MUSEUM REKISHI TO MIZEBE NO FUREAI NO MORI
b : Fukuoka-shi, Fukuoka
c : Fukuoka-City
d : Fukuoka-shi, Axs Satow Inc.
e : Axs Satow Inc.Iwao Jiki Tile Design Inc.
f : Shimizu, Taisho J.V.
h : Flooring express warmth by rugged material feeling.Flooring's durability is preeminent because natural materials was processed a high temperature calcination.
i : Permeability porous vitreous
j : Iwao Jiki Kogyo Co.,Ltd.

a：乃村工藝社大阪事業所III
b：大阪府大阪市
c：乃村工藝社
d：蔭山利夫，郷力憲治
e：乃村一級建築士事務所，建築委員会
f：乃村工藝社，建築メディア事業部，明和工務店
g：大東正巳
h：ホスピタリティを重視した構成で，心から「ようこそ」を表現。来訪の目的に合わせて，おもてなしの場を演出したい。
i：天井／ジブトーン・キューブ，EP 塗装，床／ウールカーペット，フローリング
j：乃村工藝社
●

a：NOMURA CO.,LTD., OSAKA OFFICE III
b：Osaka-shi, Osaka
c：Nomura Co.,Ltd.
d：Toshio Kageyama, Kenji Goriki
e：Nomura Architecture Office, Architecture Committee
f：Nomura Co.,Ltd., Architecture Department, Meiwa Co.,Ltd.
g：Masami Daito
h：A warm feeling of hospitality has been emphasized in the design of this construction. Various receptions have been produced to match the purpose of the visit.
i：Ceiling/jibtone cubes, EP paint, Floor/wool carpeting, flooring
j：Nomura Co.,Ltd.

a：佐賀市総合文化会館スモーキングスタンド，ダストボックス
b：佐賀県佐賀市
c：佐賀市
d：東畑・石橋建築 J.V.
e：岩尾磁器工業
h：青呉須を使ったデザインで有田焼の特色を表現した。
i：磁器
j：岩尾磁器工業
●
a：SAGA CULTURE HOLL.SMOKING-STANDO AND DUSTBOX
b：Saga-shi, Saga
c：Saga-City
d：Tohata, Ishibashi J.V.
e：Iwao jiki Kogyo Co.,Ltd.
h：Interior furniture made by this kind of ceramics are designed with using Aogosu and express the feature of Arita ware.
i：Ceramics
j：Iwao Jiki Kogyo Co.,Ltd.

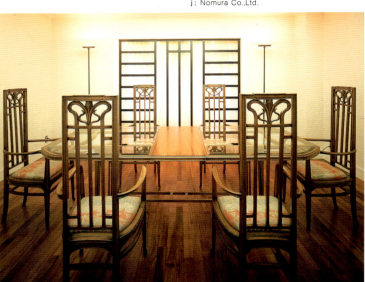

a：乃村工藝社大阪事業所II役員室
b：大阪府大阪市
c：乃村工藝社
d：蔭山利夫，郷力憲治
e：乃村一級建築士事務所，建築委員会
f：乃村工藝社，建築メディア事業部，明和工務店
g：大東正巳
h：イス1脚の造形にも気を配り，職場環境はなによりクリエイティビティを大切にをモットーとした。
i：床／ウールカーペット，天井／寒冷沙，EP 塗装仕上げ，ワイヤーフレーム，スクリーン
j：乃村工藝社
●

a：NTT POCKET PARK ねやがわ
b：大阪府寝屋川市
c：NTT 寝屋川支店
d：池上俊郎
e：池上俊郎，森井浩一，上野美子
f：建築：共立建設／内装：大丸装工事業部
g：アーバンガウス研究所
h：特徴のなかった従来の建物を，地域住民や沿道を行く車にも楽しみを与えるポケットパークに変換した。
i：ステンレス／強化ガラス／ポリカーボネート／スチール
j：池上俊郎＋アーバンガウス研究所
●
a：NTT POCKET PARK NEYAGAWA
b：Neyagawa-shi, Osaka

a：NOMURA CO.,LTD., OSAKA OFFICE II
b：Osaka-shi, Osaka
c：Nomura Co.,Ltd.
d：Toshio Kageyama, Kenji Goriki
e：Nomura Architecture Office, Architecture Committee
f：Nomura Co.,Ltd., Architecture Department, Meiwa Co.,Ltd.
g：Masami Daito
h：Great care was inserted in the construction of each chair.A creative side of the design has been emphasized to produce a working environment more than anything else.
i：Floor/wool carpeting, Ceiling/sand, EP paint, wire-frame screen
j：Nomura Co.,Ltd.

c：NTT Neyagawa Branch
d：Toshiroh Ikegami
e：Toshiroh Ikegami, Yoshikazu Morii, Yoshiko Ueno
f：Kyoritsu Construction Company/Daimaru Design and Engineering
g：Urban Gauss
h：A conventional featureless building is changed into a pocket park which give a delight view to local residents and car drivers which pass through the road.
i：Stainless-steel, toughened glass, polycarbonate, steel
j：Toshiroh Ikegami+Urban Gauss

Chapter 4

ランドスケープデザイン
Landscape Design

地域社会との融合をめざす創造拠点
かながわサイエンスパーク

株式会社日本設計

「環境」としてのかながわサイエンスパークには，創造的な人々が集い，科学技術の先端の分野で，創造的で21世紀の日本の社会に真に価値ある「知」を生み，情報を発信する「場」にふさわしいものとなることが求められました。日進月歩する研究開発の変化のなかで，人間にとってやさしい「知」が芽ばえ育つためには，人と人，人と社会，人と時間，人と自然との豊かな対話がスムーズに生まれ，研究者の向上心とインスピレーションを刺激する環境となることが望まれます。また一方で，大規模な都市施設として，このサイエンスパークが創りだす環境は，周辺地域の新しい街づくりの歴史のなかで，人々の印象に残る風景となるものでなければなりません。

かながわサイエンスパークは，広域での都市軸ともいえる第三京浜とJR南武線が交差する付近に位置しています。高層で都市的スケールをもつ，十字型の研究開発型企業ビルは，この都市軸にあわせ都市のランドマークとなるよう計画しました。店舗，ホテル，ギャラリー，多目的ホール等のある中層のイノベーションセンタービルは，住民に開かれたコミュニティーの核のひとつとして，周辺の建物や道路の軸にあわせて配置し，45度回転した二つの軸によって

できる正方形グリットをオーバーラップさせて，二つの建物を一体化させています。そして，それぞれの軸の交点となる場に意味をもたせ，モニュメントや環境装置を置いて全体の環境計画にリズムをもたせるように計画しています。

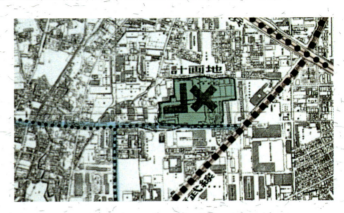

A Creative Base point Aimed at Integration with the Local Community   The Kanagawa Science Park

Nihon Sekkei Ltd.

Using the ENVIRONMENT as a theme, the Kanagawa Science Park was constructed to attract many creative people in order to produce a certain KNOWLEGE that they will become truly valuable elements of Japanese society in the 21st century as well as to offer a place for information gathering.

In the midst of rapidly-advancing research development, the onus was placed upon creating an environment in which people could easily share rich communications with each other, society, time and nature, and the ambition and inspiration of would-be researchers could be stimulated in order get the gentle feeling of the KNOWLEGE off the ground. Also, the science park could serve as an environmental view that would remain in peoples' minds as a part of the history of this large scale urban facility.

The Kanagawa Science Park is located in the vicinity of the Keihin National Highway No.3, that large axix where the huge urban area and the Japan Railways Nambu Line meet. This urban area of high-rise buildings was designed to serve as a landmark of the city in conjuction with the axix. The mediumsized Inovation Center contains shops, a hotel, a gallery and multi-purpose halls, etc., and was designed to serve as a community center for the local people. It is located on the axix of the surrounding buildings and road, and the two buildings have been joioned into one by overlapping the two square grids formed by turning both axix' to an angle of 45 degrees. The other elements, such as the monument and the environmental device, were placed in locations where each axix met in order to give meaning and rhythm to the entire environmental plan.

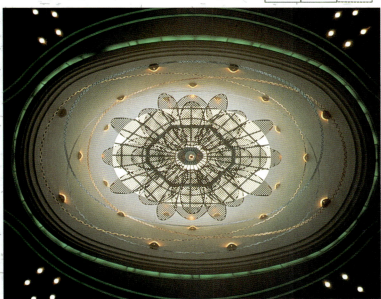

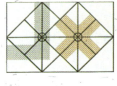

水　　　　　　　　　　　　　　　Water

人と人，人と自然との豊かな対話が，研究者の五感を通してスムーズに
芽生える，創造とやすらぎに満ちた場となるよう，清涼感と開放感それに落着き
を与えてくれる水面，水の流れ，波紋，四季の変化に富む雑木林，
そこに集う野鳥等々，願いをこめて。

The water's surface, the flow, the ripples, the thicket of assorted trees for each season, and the wild birds were all part of a design to offer freshness, openness and peace of mind in order to produce a place rich in creativity and serenity where people can share fruitful communication with each other and nature, and the inspiration of researchers could be stimulated by the constant use of all four senses.

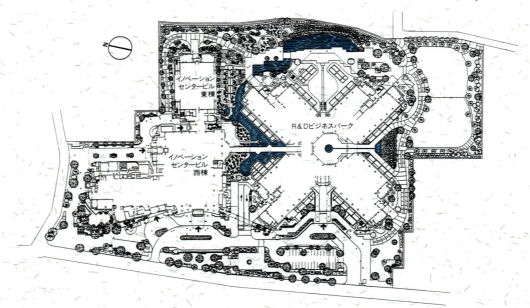

建物群を敷地境界からできるだけ離して配置し，敷地外縁に建物の大きな
ボリュームに均衡する，まとまった奥行のある緑のオープンスペースがとれるよう
努めました。この開放された緑地空間は，かながわサイエンスパークが
単に研究開発型企業の集積の場ではなく，創造活動の拠点として，また
地域のコミュニティの熟成を願う施設であることを示唆する環境を形成しています。

The volumous grouping of the buildings were purposely placed well away from the boundaries of the site to maintain the balance with the greenery when viewed from without. A large green and open space was deliberately secured. This area suggests that Kanagawa Science Park is a facility to be used not only for research and development companies, but also as a base for creative activities and a place in which local communications can nature.

光   Light

イノベーションセンタービル，研究開発型企業ビルの中心を通り南北にのびる
この中心軸は，かながわサイエンスパーク全体の建築空間の骨格となって
います。この軸線上の空間のテーマは「浮遊感」と「地，水，火，
風と空(光)」です。風のゲートに始まり灯の道，光の天井，水面にかかる渡廊下，
浮遊する光体，大地の道へと続くデザインコンセプトは，
かながわサイエンスパークの環境計画を象徴しています。

This central axix stretching from south to north through the center of
the Innovation and Research Development Company building serves
as a framework for the entire complex of the Kanagawa Sience Park.
The themes of this space on the axix line are Floatation, Earth, Water,
Fire, Wind and Sky(Light). The themes start with the Wind Gate, the
Light Ceiling, the Lighting Path – a connecting corridor suspended
over the water – and the Floating Light Body, and end with the path
to earth. This design concept symbolizes the environmental plan of the
Kanagawa Sience Park.

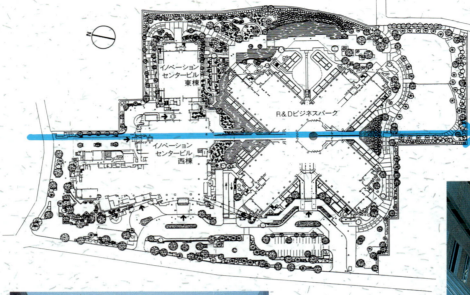

代表者＝株式会社日本設計
所在地＝神奈川県川崎市高津区坂戸
クライアント＝飛島建設株式会社開発事業部
ディレクター＝斎藤繁喜，小林正一，四手井孝樹
デザイナー＝森林兵衛，加藤友義，近宮健一
協力設計者＝株式会社愛植物設計事務所，有限会
社美術計画工房
施工者＝飛島建設株式会社横浜支店
撮影者＝川澄建築写真事務所，銀総，廣田治雄，
翠光社

Applicant=Nihon Sekkei Inc.
Site=Sakato, Takatsu - ku, Kawasaki - shi,
Kanagawa
Client=Tobishima Corporation Development
Operation Div.
Director=Shigeki Saito, Shoichi Kobayashi,
Takaki Shidei
Designer=Rinbei Mori, Tomoyoshi Kato, Ken'
ichi Chikamiya
Design Cooperater=Art Planning Studio Co.,
Ai-Shokubutsu Landscape Planning Office
Contractor=Tobishima Corporation Yoko-
hama Branch
Photographer=Kawasumi Architectural Photo-
graph Office, Ginso, Haruo Hirota, Suikosha

a：福岡タワー
b：福岡県福岡市
c：福岡タワー
d：山本博
e：藤田和孝，西田正則
g：西日本写房
h：沿道と海浜公園をつなぐ福岡タワーのシンボル性を強調する舗装パターンをもったイベント広場。
i：インターロッキングブロック舗装，クスノキ，アイビー
j：日建設計
●
a：FUKUOKA TOWER
b：Fukuoka-shi, Fukuoka
c：Fukuoka Tower Co.,Ltd.
d：Hiroshi Yamamoto
e：Kazunori Fujita, Masanori Nishida
g：Nishinihon Shabo
h：This is an event plaza with a paving pattern that emphasizes the symbolism of the tower connected to the roadside and seaside park.
i：Inter-locking block paving, camphor trees, ivy
j：Nikken Sekkei Ltd.

a：心斎橋商店街アーケード
b：大阪府大阪市
c：心斎橋筋商店街振興組合
e：竹中工務店設計部/狩野忠正, 井上宜延, 峠孝治
f：竹中工務店
g：竹中工務店/古川泰造
h：吊橋方式のアーケードで，天井は花びらをイメージ
し，中央部は左右に開閉する。
i：鉄骨, アルミキャスト, ポリカーボネート板, 花こう岩
j：竹中工務店
●
a : SHINSA BASHI SHOPPING MALL ARCADE
b : Osaka-shi, Osaka
c : Shinsaibashi Shopping Mall Promotion
Association
e : Takenaka Corporation, Building Design
Dep./Tadamasa Kano, Yoshinobu Inoue,
Takaharu Toge
f : Takenaka Corporation
g : Takenaka Corporation/Taizo Furukawa
h : This was designed as a supension bridge-
style arcade with the ceiling in the image of
flower petals.The center part opens on both side.
i : Reinforcement, aluminium cast,
polycarbonate boards, granite veneer
j : Takenaka Corporation

a：魅力ある道路づくり事業 区役所前通り
b：神奈川県横浜市
c：横浜市道路局
d：横浜市都市計画局
e：西澤健，森田昌嗣，磯村克郎
f：前田・山岸・菊岡建設共同企業体
g：T.ナカサ・アンド・パートナーズ
h：道路に地形を構築すること，路上の行為の場を設定すること，重厚な基盤に軽い構造を作ること。
i：白御影石，御影質磁器タイル，ステンレス，耐候性鋼材
j：GK 設計
●
a：KUYAKUSHO-DORI AVENUE
b：Yokohama-shi, Kanagawa
c：Road & Highway Bureau, Yokohama City
d：Urban Planning Bureau, Yokohama City
e：Takeshi Nishizawa, Yoshitsugu Morita, Katsuro Isomura
f：Maeda Yamagishi Kikuoka Construction J.V.
g：T.Nacasa & Partners
h：The aims of this road construction are to construct a configuration of the ground upon its surface, to create a place of action on the road and to build a light construction on a heavy foundation.
i：White granite stone, granite temperament ceramic tiles, stainless steel, all-weather steel material
j：GK Sekkei Inc.

a：加古川駅前通商店街モール
b：兵庫県加古川市
c：加古川駅前通商店街振興組合
e：稲地一晃
f：川重運輸建設，大林道路
g：稲地一晃
h：老若男女が安心して買物や，散策の楽しみが味わえる買物広場的なモールの実現を目指した。
i：御影石，カラーアルミ，ポリカーボネイト，スチール，タイル
j：計画工房 DNA
●
a：KAKOGAWA EKIMAE-DORI SHOTENGAI MALL
b：Kakogawa-shi, Hyogo
c：Kakogawa Ekimae-dori Shotengai Shinkokumiai
e：Kazuaki Inachi
f：Kawaju Unyu Kensetsu, Obayashi Douro
g：Kazuaki Inachi
h：It is designed to actualize a mall like a shopping park where men and women of all ages can enjoy shopping and walking feeling at rest.
i：Granite, color aluminum, polycarbonate, steel, tile
j：Keikaku Kobo DNA

a：久屋大通公園　復興事業収束モニュメント
b：愛知県名古屋市
c：名古屋市
d：日建設計／南石周作
e：日建設計／市川国夫，若林亮
f：安藤・鈴中 J.V.
g：SS名古屋
h：未来へ向かう緑の川（公園）の船。レーザー光は1km
先のテレビ塔を結び，夜空に新たな都市景観を創る。
i：モニュメント：ステンレス
j：日建設計

●

a：HISAYA ODORI PARK-MONUMENT
b：Nagoya-shi, Aichi
c：Nagoya City
d：Nikken Sekkei/Shusaku Nanseki
e：Nikken Sekkei/Kunio Ichikawa, Makoto
Wakabayashi
f：Ando & Suzunaka, J.V.
g：SS Nagoya
h：A green river (park) ship heading for the
future. The laser beam creates a new urban
view in the sky by being connected to the
television broadcasting tower located one
kilomenter distant.
i：Monument : stainless steel
j：Nikken Sekkei Ltd.

a：レイトン・ハウス　横浜
b：神奈川県横浜市
c：レイント
d：江幡正之
e：ランドスケープデザイン：豊田幸夫, 彫刻：児玉慎憲
f：鹿島建設
g：川澄建築写真事務所, 清水写真事務所：清水昭
h：昼と夜の景観を考え, モニュメントとしてのゲート,
風で動くオブジェと光のオブジェを配置した。
i：ステンレス, 御影石, 磁器タイル
j：鹿島建設
●
a：LYETON HOUSE YOKOHAMA
b：Yokohama-shi, Kanagawa
c：Lyeton Co.,Ltd.
d：Masao Ebata
e：Landscape Design : Yukio Toyoda,
Sculpture & Monument : Shinken Kodama
f：Kajima Corporation
g：Kawasumi Architectural Photo Office,
Shimizu Photo Office : Akira Shimizu
h：Three elements-a symbolic gate, a wind
sculpture and a light ornament-have been
used to enhance the landscape in both
daylight and at night.
i：Stainless steel, granite veneer, porcelain tiles
j：Kajima Corporation

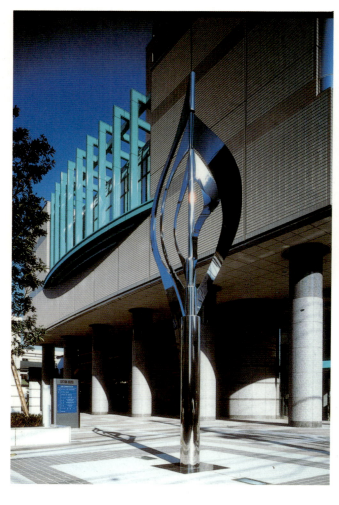

a：トキメックビル
b：東京都大田区
c：トキメック
d：鹿島建設／飯笹正勝
e：鹿島建設／小野寺康夫，彫刻：児玉慎憲，児玉もえみ
f：鹿島建設
g：川澄建築写真事務所，クリエイティブ・スタジオ・カ
ワカミ，SS 東京
h：トキメックの工場跡地の再開発計画。公開空地，緑
を多く取り入れ，周囲の環境改善に大きく貢献。
i：コンクリートブロック，タイル，ドイツトウヒ，シラカシ，他
j：鹿島建設
●

a: TOKIMEC BUILDING
b: Ota-ku, Tokyo
c: Tokimec Inc.
d: Kajima Corporation/Masakatsu Iizasa
e: Kajima Corporation/Yasuo Onodera,
Sculptor : Shinken Kodama, Moemi Kodama
f: Kajima Corporation
g: Kawasumi Architectural Photograph Office,
Creative Studio Kawakami, SS Tokyo Ltd.
h: A redevelopment plan for the old site of the
Tokimec factory. An open sapce. With the
adoption of much greenery, a great
contribution has been paid to the improvement
of the surrounding environment.
i: Concrete blocks, tiles, German spruce,
white oak, etc.
j: Kajima Corporation

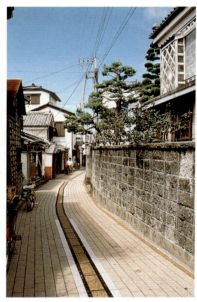

a：西伊豆町総合ギャラリーパーク計画
b：静岡県加茂郡
c：西伊豆町
d：西澤健，朝倉則幸，花輪恒
e：松永博巳，亀谷美幸，白浜力，田村賢治，佐々木賢範
f：堤工務店，藤井組
g：安川千秋，T.ナカサ・アンド・パートナーズ，斉藤さだむ
h：町にもともとある特徴をデザインに生かした広場の
ネットワークで，町を再確認させることを意図している。
i：浜辺のゴロタ石，漁船のマスト，漁網
j：GK 設計
●
a：TOTAL ENVIRONMENT DESIGN PROJECT
OF NISHI-IZU-CHO
b：Kamo-gun, Shizuoka
c：Nishi-Izu-cho.
d：Takeshi Nishizawa, Noriyuki Asakura,
Hisashi Hanawa
e：Hiromi Matsunaga, Miyuki Kameya, Tsutomu
Shirahama, Kenji Tamura, Takanori Sasaki
f：Tsutsumi Komuten Co.,Ltd., Fujii-gumi Co.,Ltd.
g：Chiaki Yasukawa, T.Nacasa & Partners,
Sadamu Saito
h：This project has the intention of making
people re-appreciate the merits of the town by
improving its existing features in design by
constructing a public netwked square.
i：Cobble stones from the beach, the mast
and net of fishing boat
j：GK Sekkei Inc.

a：名古屋商科大学コミュニティーセンター
b：愛知県愛知郡
c：栗本学園
e：竹中工務店設計部/古田博司，西本博，冨田徹
f：竹中工務店
g：村井修
h：緑豊かな郊外型大学にキャンパスライフのコアとなる「街ープロムナード」を創る。
i：Pca コンクリート，熱線反射ガラス，タイル
j：竹中工務店
●

a : NAGOYA UNIVERSITY OF COMMERCE AND BUSINESS ADMINISTRATION, COMMUNITY CENTER
b : Aichi-gun, Aichi
c : Kurimoto Gakuen
e : Takenaka Corporation, Building Design Dep./Hiroshi Furuta, Hiroshi Nishimoto, Toru Tomita
f : Takenaka Corporation
g : Osamu Murai
h : A Street Promenade, which becomes the core of campus life in an urban-style university rich in greenery, has been created here.
i : Pca concrete, heat-reflecting glass, tiles
j : Takenaka Corporation

a：南大津通商店街モール　ピュア O2
b：愛知県名古屋市
c：南大津通商店街振興組合
e：竹中工務店設計部／三井富雄，繁野舜，集山一廣，
河崎康了
f：竹中工務店
g：フォトハウス田中／田中昌彦，車田写真事務所／車田保
h：「歴史と現代文化を感じる街へ」というテーマのも
とに，光と水と緑で豊かに街を演出する。
i：床：200角タイル／ストリートファニチャー：ステンレ
ス，御影石，レッドウッド
j：竹中工務店
●

a：MINAMI-OTSU SHOPPING MALL-PIER O2
b：Nagoya-shi, Aichi
c：Minami-Otsu Shopping District Promotive
Association
e：Takenaka Corporation, Building Design
Dep./Tomio Mitsui, Hajime Shigeno, Kazuhiro
Shuyama, Yasunori Kawasaki
f：Takenaka Corporation
g：Photo House Tanaka/Masahiko Tanaka,
Kurumada Photograph Office/Tamotsu Kurumada
h：Full quantities of light, water and greenery
were used to produce this street, which is
based on the theme CREATION OF A STREET
EXUDING HISTORY AND MODERN CULTURE.
i：Foolr : 200 sq.Tiles/Street furniture :
stainless steel, granite stone, redwood
j：Takenaka Corporation

a：西池袋公園
b：東京都豊島区
c：豊島区公園緑地課
d：上村彰雄
e：長山秀明，小沢真由美，高瀬昭男
f：西武造園
g：長山秀明，上村彰雄
h：暗く不人気になっていた公園を，都市の明るいオア
シスとして再生した。
i：花こう岩，擬石ブロック
j：アーク造園設計事務所
●

a：NISHI-IKEBUKURO PARK
b：Toshima-ku, Tokyo
c：Toshima-ku Koen Ryokuchi-ka
d：Akio Kamimura
e：Hideaki Kaiyama, Mayumi Ozawa, Akio Takase
f：Seibu Zoen
g：Hideaki Kaiyama, Akio Kamimura
h：This park had lost its popularity since it was
dark.It came to life again as a pleasant oasis
in a city.
i：Granite, imitation stone block
j：URC Landscape Design Office

a：わんぱく天国
b：東京都墨田区
c：墨田区公園河川課，社会教育課
e：貝山秀明，小野里宏
f：柳島寿々喜園
g：貝山秀明
h：広場を失った下町の子供たちのための，自分たち
の責任でなんでもできる自由な公園。
i：杉材
j：アーク造園設計事務所
●

a：WANPAKU-TENGOKU
b：Sumida-ku, Tokyo
c：Sumida-ku, Koen Kasen-ka, Shakaikyoiku-ka
e：Hideaki Kaiyama, Hiroshi Onozato
f：Yanagizima Suzukien
g：Hideaki Kaiyama
h：This is an open park for Shitamachi (the
traditional shopping and entertainment districts
of Tokyo) children who have lost open places
to play everythi~ng on their own responsibility.
i：Japan cedar
j：URC Landscape Design Office

a：花園東公園
b：東京都新宿区
c：新宿区公園課
d：折口紀子
e：貝山秀明，影山温子
g：貝山秀明
h：繁華街の片隅に取り残された古い小公園を，街角
のカフェテリヤとして再生した。
i：H鋼，RC柱，ブロック舗装，モウソウ竹，黒松仕
立物
j：アーク造園設計事務所
●

a：HANAZONO HIGASHI PARK
b：Shinjuku-ku, Tokyo
c：Shinjuku-ku, Koen-ka
d：Noriko Origuchi
e：Hideaki Kaiyama, Atsuko Kageyama
g：Hideaki Kaiyama
h：An out-of-date small park in the corner of
an amusement quarter is revived as a cozy
cafeteria in a city.
i：H-section, RC post, Block pavement,
species of thick-stemmed bamboo, Japanese
black pine
j：URC Landscape Design Office

a：ホテル日航 福岡
b：福岡県福岡市
c：九州勧業，ホテル日航福岡
d：平田翰那
e：ランドスケープデザイン：黒田幸夫，照明コンサル
タント：J.フィッシャー，ポール・マランツ，TL ヤマギワ研究所
f：鹿島建設
g：川澄建築写真事務所
h：ステンレスと砂利，そして照明により竜安寺の石庭
を現代の石庭として作庭。
i：ステンレス，白那智石，黒那智石，玄晶石
j：鹿島建設
●
a：HOTEL NIKKO FUKUOKA
b：Fukuoka-shi, Fukuoka
c：Kyushu Kangyo, Hotel Nikko Fukuoka
d：Kanna Hirata
e：Landscape Design : Yukio Kuroda, Lighting
Concerting : Jules Fisher, Paul Marantz, TL
Yamagiwa Laboratory Inc.
f：Kajima Corporation
g：Kawasumi Architectural Photo Office
h：A modern version of Ryuan-ji's stone
garden using stainless steel, gravel and
recessed lighting.
i：Stainless steel, white Nachi stones, black
Nachi stones, Gensho stones
j：Kajima Corporation

a：デナマボール
b：東京都足立区
c：足立区役所
d：須藤
e：イデア造園設計
f：サカエ
g：井上和幸
h：不思議な球体は昼と夜の顔をもっている。夜になる
とイルミネーションが小宇宙をメタファする。
i：モザイクタイル，LED
j：サカエ
●
a：DENAMA-BALL
b：Adachi-ku, Tokyo
c：Adachi Ward Office
d：Sudo
e：Idea Zoen Sekkei
f：Sakae Co.,Ltd.
g：Kazuyuki Inoue
h：A wonder ball has two kind of faces.One is
for the daytime, another is for the nighttime.
When the night fells it's illumination makes a
metaphoricalmicrocosmos.
i：Mosaic tile, LED
j：Sakae Co.,Ltd.

Chapter 5

住環境デザイン
Residential Environmental Design

スペースとパターンが織りなす住空間ハーモニー
CITY SCREEN XIII 岸田邸

池上俊郎＋アーバンガウス研究所

兵庫県西宮市に立地する。近傍には有名私学があり，文教地域である。各区画は，400平方メートルを越える良好な住宅地である。建物は，若夫婦と母親のための木造二階建ての二世帯住居である。建物は，平面的には，
(1) 9m×9m の正方形，
(2) 5.4m×16.8m の長方形が，
(3) 2.7m×2.7m の正方形によって結ばれた形式をもつ。

それらの関係性に，(4) 集成材の列柱が配置されることによって，緊張感をもつ内外空間が形成されている。
これらの平面は互いに，3.75度，7.5度のずれをもたされて配されている。
(1) は若夫婦の住居，(2) は母親の住居，(3) は(1)(2)を結ぶ母親の玄関として設定されている。

若夫婦の住居は，正方形プランの内部に「コムウッド」と呼ばれるスウェーデン製の12角形の集成材の柱が配置されている。直径300mmを使用した。平面の中心には，4.5m×4.5mの正方形グリッドのシンボリックスペースを設けた。二階には，∪字形の回廊があり，変形四角錐の屋上出口に到る階段が設けられている。回遊性と垂直方向の浮遊感の獲得を通じて，精神的中心スペースの形成を図った。「コムウッド」と呼ばれる集成材の柱は，ここでは空間上極めて大きな役割をもっている。日本の木造の建物のスケールを越えた直径300mmの柱は，内外にわたって建築全体を精神的にも，物理的にも支配している。母親の住居は，100年の歴史を内蔵する仏壇を納める和室，四角錐の吹抜けをもつ居間等から成立する。

二つの棟を結ぶスペースは，一階では母親の玄関であり，二階では物干しの役割をもつテラスである。奥まった位置に配されたこのスペースは，建物全体における機能の変換装置である。また，庭にアルコーブをうみ，屋外空間に広がりを与えている。

こうした正方形をベースとする相互の関係性は，ペーブメントのパターンとして，庭においても探られている。円弧状の床パターンはこうした応答を強調している。東南側よりのアプローチは，ゆるやかな勾配と，平面上のカーブをもっている。2枚の壁は，それぞれ高さを異にし，視覚的差異を生み，庭をステージアーキテクチャーとして活性化する。屋上に設けられた変形四角錐は，都市空間スケールにおいて，街の方向性と記号性を探り，建築内部においては，居住空間に豊かな午後の光を取り入れている。また，オーナーの瞑想の場を形成する役割をも果たしている。新たな空間体験をともなう，本住宅のシンボルとなっている。本建築は，外部にグレイを中心とする4段階の色彩構成がなされている。パールグレイ，フレンチグレイ，ダークグレイである。これらは，序列をもつことによって，視覚的世界に強弱を生み，奥行と広がりを形成している。全体として，内外空間における開放系を目指し，シンボル性に満ちた空間体験を生み出せたらと考えている。

Harmony of interior space with pavement pattern
CITY SCREEN XIII KISHIDA R.

Toshiroh Ikegami + Urban Gauss

Located in Nishinomiya city, Hyogo prefecture. An area well known for its education with a famous private school nearby. This is an excellent residential area consisting of many land lost exceeding 400 square meters in size. This is a two storied wooden house built to accomodate two seperate families; a young couple and their mother. The ground floor is made of (1) a square measuring 9m×9m, and (2) a rectangular shape measuring 5.4m×16.8m, and these are linked to (3) the square shaped land that measures 2.7m×2.7m. The assembly colonnade was especially arranged in tis linkage sections to produce a space with tension. The floor plans were arranged to overlap each by an angle of 3.75m×7.5m. Picture (1) is the living quarters for the young couple, and picture (2) is for their mother. Picture (3) has been created to serve as an entrance for the mother that links the space of both (1) and (2).

The young couple's living space consists of a square plan with Swedish style 12-side assembly columns known as Comwood arranged within. Columns with a diamerter of 300mm have been used. A symbolic grid area in which a 4.5m×4.5m square has been created in the center of the 1st floor. The second floor has a U-shaped corridor and the stairs lead to the entrance of the transformational cubic-shaped rooftop. By gaining the feel of a circular tour and virtical directional floatation, a spritual based space has been created. The Comwood columns play an extermely important role in this space. They have a far larger diameter that those used in most Japanese houses, and therefore dominate the entirebuilding both spiritually and physically.

The mother's living quarters is based on the Japanese style with a built-in family Buddhist alter in which 100 years of the family's history in placed. The living space itself is a cubic void. The space linking the two homes together is the entrance to the mother's living quarters on the 1st floor and terrace on which washing can be hung. This innermost arranged space serves as a functional alternating device for the entire building. It also forms an alcove in the garden with appears to expand the space available outdoors.

A similar mutual connection of square shapes is also found in the patterns of the paving stones in the garden, and the circular arc-shaped pattern of flooring enhances such a respense.

The approach from the south-east side is gently sloped and has a plainer curve. A visual contrast has been achieved by constructing one wall higher than the other in order to activate the garden as a stage for the architecture.

The transformational cubic shape on the rooftop offers a feeling of directionality and registration in the street along the scale of urbaninity, as well as taking the rich afternoon sun into the living space. It serves as a place where the owner can be lost in meditation and is a symbol of the house involving the experience of new space. The external surface of the building is composed of four different colors based on grey; Pearl grey, French grey, grey and dark grey. These colors produce a strong contrast in the visual world and form a depth and wideness in order of importance. On the whole, my aim was to create a open space both within and without and to produce a space rich in symbolism.

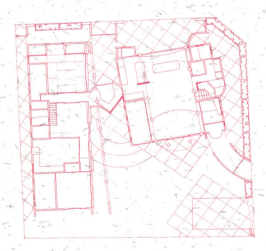
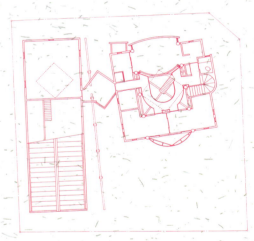

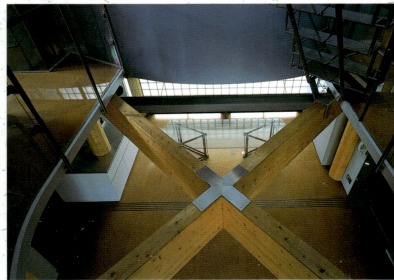

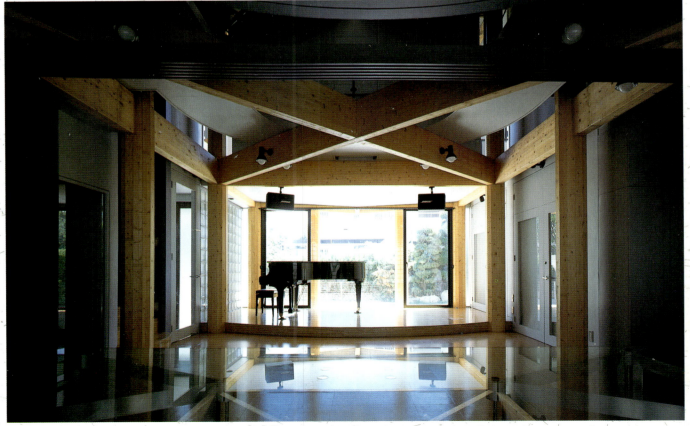

作品名=CITY SCREEN XIII 岸田邸
所在地=兵庫県西宮市
クライアント=岸田隆之
ディレクター=池上俊郎
デザイナー=池上俊郎+アーバンガウス研究所 担当,
山家弘
施工者=リホミカ
撮影者=松村芳治
主な材料=外壁:ラスモルタル下地アクソナイト塗布/
柱:スウェーデン製パイン集成材(コムウッド)
応募代表者=池上俊郎+アーバンガウス研究所

Project name=CITY SCREEN XIII KISHIDA R.
Site=Nishinomiya-shi, Hyogo
Client=Takayuki Kishida
Director=Toshiroh Ikegami
Designer=Toshiroh Ikegami+Urban Gauss,
Hiroshi Yamaga
Contractor=Reformica
Photographer=Yoshiharu Matsumura
Principli materials=Exterior wall: Lath mortar
setting bed axonite application. A column:
Swedish pine glue-laminated timber
(comwood)
Applicant=Toshiroh Ikegami+Urban Gauss

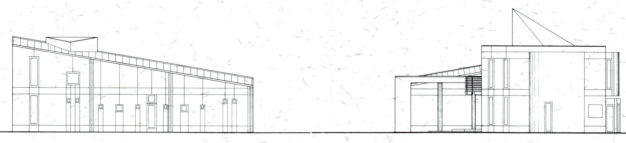

a：日立基礎研究所 美山クラブ
b：埼玉県鳩山市
c：日立基礎研究所
d：日立建設
e：鹿島建設/小野寺康夫, 日立建設
f：鹿島建設
g：SS東京
h：社員用の厚生施設。宿泊, 食事のできる和風建
築。落ち着いた雰囲気のなかでの新しいイメージの石庭。
i：芦野石, サビ砂利, シャラ, ヤマモミジ, サルスベリ, 他
j：鹿島建設
●
a：HITACHI ADVANCED RESEARCH
LABORATORY-MIYAMA CLUB
b：Hatoyama-shi, Saitama
c：Advanced Research Laboratory, Hitachi Ltd.
d：Hitachi Architects & Engineers Co.,Ltd.
e：Kajima Corporation/Yasuo Onodera,
Hitachi Architects & Engineers Co.,Ltd.
f：Kajima Corporation
g：SS Tokyo
h：A welfare facility for wirklers' use only. This
Japanese-style building offers accomodation
and meals. It has a new-image stone garden in
a quiet atmosphere.
i：Ashino stone, rusty gravel, shalla, mountain
maple, indian lilac, etc.
j：Kajima Corporation

a：ロワ・ヴェール荻窪北
b：東京都杉並区
c：東京建物
e：大成建設設計本部/山口一平, 今里清, 正木素
行, 中村和久, 黒田信行
f：大成建設
g：服部寿徳（ハットリスタジオ）
h：都心としては, のんびりとした風情を大事にし, 50代
以上の年齢層をターゲットとした分譲住宅。
i：磁器タイル
j：大成建設設計本部
●
a：ROI VERT OGIKUBO-KITA
b：Suginami-ku, Tokyo
c：Tokyo Tatemono Corp.
e：Taisei Corporation Design and Proposal
Division, Kunihira Yamaguchi, Kiyoshi Imasato,
Motoyuki Masaki, Kazuhisa Nakamura,
Nobuyuki Kuroda
f：Taisei Corporation
g：Toshinori Hattori(Hattori Studio)
h：A house built for sale is designed to value
leisure atmosphere is spite of the midtown
area and aim at people who are more than
fifties.
i：Vitreoustile
j：Taisei Corporation Design and Proposal
Division

a：池田山ゲストハウス
b：東京都品川区
c：潮田洋一郎
e：竹中工務店設計部/小嶋敏夫，地主道夫，澤田勝
f：竹中工務店
g：マツダ・プロ・カラー
h：中庭に面したオープンな開口を設け，自然環境
（光，風，緑）を内部空間へ取込む手法とした。
i：シンボルツリー，花こう岩，アルミパネル，ガラス，
ステンレス
j：竹中工務店
●
a：IKEDAYAMA GEST HOUSE
b：Shinagawa-ku, Tokyo
c：Yoichiro Shioda
e：Takenaka Corporation, Building Design
Dep./Toshio Kojima, Michio Jinushi, Masaru
Sawada
f：Takenaka Corporation
g：Matsuda Pro Color
h：The natural environment (light, wind and
greenery) has been introduced within this
building by creating a free opening that faces
onto the courtyard.
i：Symbol trees, granite veneer, aluminium
panel, glass, stainless steel
j：Takenaka Corporation

a：SHIRAKAWA RESIDENCE
b：兵庫県神戸市
c：白川地所
e：大林組本店設計部，三谷幸司，堀口壮平
f：大林組
g：ナトリ光房
h：保存建物を新築建物とともに閑静な住環境を創り出すしかけとして扱い，周辺環境との調和を図った。
i：磁器二丁掛タイル，花こう石，アルミ，吹き付けタイル，ステンレス
j：大林組
●

a：SHIRAKAWA RESIDENCE
b：Kobe-shi, Hyogo
c：Shirakawa Jisho Co.,Ltd.
e：Obayashi Corp., Design Section : Koji Mitani, Sohei Horiguchi
f：Obayashi Corporation
g：Studio Natori
h：A reserve building is utilized to create a quiet housing environment coexist with new-built building in order to harmonize with surroundings.
i：Double sized vitreous tiles, granite, aluminum, sprayed tile, stainless steel
j：Obayashi Corporation

a：駒沢ガーデンハウス
b：東京都世田谷区
c：三菱地所
e：竹中工務店設計部/金沢義人，南裕二郎
f：竹中工務店
g：マキ・フォトグラフィック/牧明夫
h：総合設計により敷地の大半を公用空地とし，既存の自然緑地を主体とした外部環境を創造。
i：既存樹林，インターロッキング，木製複合遊具，レンガタイル，花こう岩，木製ベンチ
j：竹中工務店
●

a：KOMAZAWA GARDEN HOUSE
b：Setagaya-ku, Tokyo
c：Mitsubishi Estate Co.,Ltd.
e：Takenaka Corporation, Building Design Dep./Yoshihito Kanazawa, Yujiro Minami
f：Takenaka Corporation
g：Maki Photographic/Akio Maki
h：With an all out design, most of the site has been built as a public open space in order to create an outside enviroment based on the existing track of natural green land as a living body ; a scarecity in most cities.
i：Existing trees, interlocking, complex wooden playthings, brick tiles, granite stone, wooden benches
j：Takenaka Corporation

a：世田谷弦巻ハイム
b：東京都世田谷区
c：住友商事
d：竹中工務店東京本店設計部/小嶋敏男
e：竹中工務店東京本店設計部/高須芳史
f：竹中工務店
g：東京グラフィック/門馬金昭
h：自然環境（光，空気，緑）を内部まで導くように個々のエレメントを動線上に計画配置した。
i：シンボルツリー，並木，花こう岩，小口タイル，ピンコロ石，ナラ練付（軒天，門，玄関扉）
j：竹中工務店

●

a：SETAGAYA TSURUMAKI HEIM
b：Setagaya-ku, Tokyo
c：Sumitomo Corporation
d：Takenaka Corporation Tokyo Branch,
Building Design Dep./Toshio Kojima
e：Takenaka Corporation Tokyo Branch,
Building Design Dep./Yoshifumi Takasu
f：Takenaka Corporation
g：Tokyo Graphic/Kaneaki Monma
h：Individual elements were specially planned
and arranged on the flow-line in order to bring
the natural environment (light, air, greenery)
into the room.
i：Symbol trees, rows of trees, granite veneer,
small tiles, pin stones, Japanese oak paste
adhesive (ceiling, gate, front entrance)
j：Takenaka Corporation

a：飯田ビル
b：東京都千代田区
c：飯田孝
e：大成建設設計本部/山口一平，橋本緑郎，佐藤誠
f：大成建設
g：三輪晃士（三輪晃久写真研究所）
h：吹き抜けを中心にくつろいだ雰囲気のクリエイティブオフィス空間。庭付の一戸建て感覚の住宅。
i：磁器タイル，アルミ，スチール，シグマルト，ナラフローリング
j：大成建設設計本部

●

a：IIDA BULDING
b：Chiyoda-ku, Tokyo
c：Takashi Iida
e：Taisei Corporation Design and Proposal
Division, Kunihira Yamaguchi, Rokuro
Hashimoto, Makoto Sato
f：Taisei Corporation
g：Koshi Miwa(Kokyu Miwa Architectural
Photograph Laboratory)
h：A building containing a creative office
space and detached house with yard type
having a comfortable atmosphere centering
around an atrium.
i：Vitreous tile, aluminum, steel, Japanese oak
flooring
j：Taisei Corporation Design and Proposal Division

a：鎌倉河合ビル
b：神奈川県鎌倉市
c：カワイスチールエステート
e：大成建設設計本部/小笠原祥仲
f：大成建設横浜支店
g：鏑木宏（カブラギ写真事務所）
h：個性ある敷地を生かし，路地状の動線は変化ある景観を与える。
i：磁器質タイル，黒御影石
j：大成建設設計本部

●

a：KAMAKURA KAWAI BUILDING
b：Kamakura-shi, Kanagawa
c：Kawai Steel Estate Co.,Ltd.
e：Taisei Corporation Design and Proposal
Division, Yoshinaka Ogasawara
f：Taisei Corporation Yokohama Branch
g：Hiroshi Kaburagi(Kaburagi Photographic Office)
h：A building is designed to make the best
use of a unique site.Lane-state flows give us
varied scenery.
i：Vitreous tile, black granite
j：Taisei Corporation Design and Proposal Division

a：六甲アイランド CITY イーストコート4番街
b：兵庫県神戸市
c：六甲アイランド開発，積水ハウス
d：日建設計/猪倉啓行
e：日建設計/北島正行，日建ハウジング/津路善巳，
田辺真知子
f：竹中工務店
g：高橋裕嗣
h：余裕のある住棟配置によって美しい街並を創出し
た理想的な集合住宅街区。
　j：日建設計
●

a: ROKKO ISLAND CITY EAST COURT
REGION 4
b: Kobe-shi, Hyogo
c: Rokko Development Co.,Ltd., Sekisui House Ltd.
d: Nikken Sekkei/Hiroyuki Inokura
e: Nikken Sekkei/Msayuki Kitajima, Nikken
Housing/Yoshimi Tsuji, Machiko Tanabe
f: Takenaka Corporation
g: Hiotsugu Takahashi
h: This is an ideal multiple-dwelling estate, the
streets of which are beatifilly lined with
spacious housing arrangements.
　j: Nikken Sekkei Ltd.

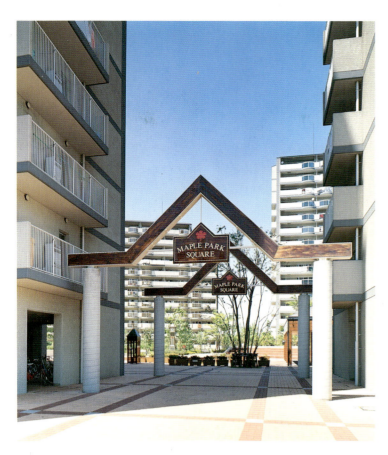

a：メイプルパークスクエア
b：大阪府守口市
c：住友不動産，東洋紡住宅，東泉開発
e：竹中工務店設計部／山下祥三，岩下龍吉
f：TMD計画共同企業体／竹中工務店，大林組，銭高組，鴻池組
g：大島勝寛
h：「住む」「遊ぶ」「集う」「学ぶ」をテーマに住棟，集会所，ランドスケープを「ブリティッシュ・カナデアン」でデザイン統一した。
i：屋根：アスファルトシングル／外壁：アクリル系スタッコ吹き付け
j：竹中工務店

a : MAPLE PARK SQUARE
b : Moriguchi-shi, Osaka
c : Sumitomo Realty & Development Co.,Ltd., Toyobo Housing Co.,Ltd., Tosen Development Co., Ltd
e : Takenaka Corporation, Building Design Dep./Shozo Yamashita, Ryukichi Iwashita
f : TMD Project J.V./Takenaka Corporation, Ohbayashi-Gumi, Zenidaka-Gumi, Konoike-Gumi
g : Katsuhiro Oshima
h : With LIVE, PLAY, ASSEMBLE and LEARN as main themes, this British Canadian design was used for the living quarters, the assembly hall and the landscape.
i : Roof : asphalt single/Exterior : acrylic stucco spray
j : Takenaka Corporation

a：グランエクレール市谷
b：東京都港区
c：M.S.D.
e：大成建設設計本部/西条由治，前田正英，五戸准
f：大成建設/堀内建設
g：岡田泰治
h：ハイグレードな都市型集合住宅。住戸タイプとともに内装仕上げ材も異なる。
i：タイル，花こう岩，大理石，アルミ，ステンレス
j：大成建設設計本部
●
a：GRAND ECLAIRER ICHIGAYA
b：Minato-ku, Tokyo
c：M.S.D.Ltd.
e：Taisei Corporation Design and Proposal Division, Yoshiharu Nishijo, Masahide Maeda, Jun Gonohe
f：Taisei Corporation, Horiuchi Corporation
g：Yasuharu Okada
h：A high-grade urban collective housing.Each dwelling unit, as well as interior finish materials, is designed individually.
i：Tiles, granite, marble, aluminum, stainless
j：Taisei Corporation Design and Proposal Division

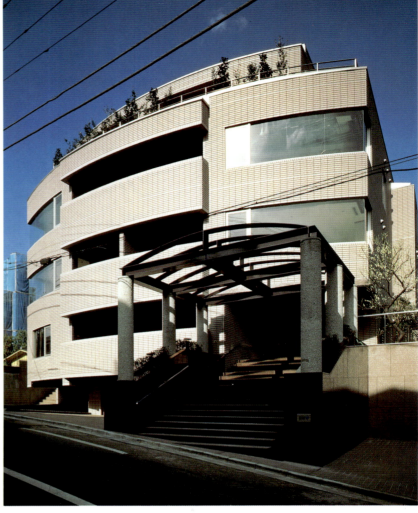

a：カデンツア・ザ・タワー モデル・ルーム
b：大阪府大阪市
c：日商岩井不動産，不二建設
d：藤山利夫
e：有居徹彦，竹本明美，倉田明夫
f：乃村工藝社
g：絹巻豊
h：人を原点とした優しさと柔らかさ。自由さと，くつろぎの実感，そしてフレキシビリティー性の高さ。
i：大理石，ウールカーペットの床，クロスと木製合板ウォール，シルクのドレープカーテン
j：乃村工藝社
●

a：CADENZA THE TOWER-MODEL ROOM
b：Osaka-shi, Osaka
c：Nissho Iwai Real Estate Co., Fuji Co.,Ltd.
d：Toshio Kageyama
e：Tetsuhiko Arii, Akemi Takemoto, Akio Kurata
f：Nomura Co.,Ltd.
g：Yutaka Kinumaki
h：A gentleness and softness based on people.An actual feel of freedom and relaxation and a high degree of flexibility.
i：Floor/marble stone & a wool carpet, walls/cloth & wood veneer board, silk drape curtains
j：Nomura Co.,Ltd.

200

a：PATIO DE ROBLE 大林組御影社宅
b：兵庫県神戸市
c：大林組
e：大林組設計部，田中弘成，栗原幸二，大塚慶子
f：大林組
g：名執一雄
h：従来の社宅のもつ画一性を排除して各住戸にアイデンティティを持たせ，コミュニティ形成を誘発する。
i：外壁：コンクリート打っ放し，モザイクタイル貼り，スタッコ吹き付け/外構：洗出し平板貼り
j：大林組
●
a：PATIO DE ROBLE
b：Kobe-shi, Hyogo
c：Obayashi Corporation
e：Obayashi Corporation, Design Departments, Hironari Tanaka, Koji Kurihara, Keiko Otsuka
f：Obayashi Corporation
g：Kazuo Natori
h：A company flat is designed to make away with uniformity which usual companyflats have make identify indivisually and induce a community.
i：Exterior wall：Exposed concrete, mosaic tile, stucco spray
j：Obayashi Corporation

a：オークヒルズ香里
b：大阪府寝屋川市
c：野村不動産，丸紅
e：(大林組)田中弘成，能丸弘毅，宮治長一，(竹中工務店)横山芳祝，吉田孝
f：大林組，竹中工務店
g：SS 大阪
h：人と車を完全に分離した動線計画とミニサンクチュアリを設けた外部空間は，人と自然のふれあいを意図している。
i：ILB舗装，舗石タイル，洗出し平板ブロック，自然石
j：大林組本店建築設計第5部
●

a : OAK HILLS KORI
b : Neyagawa-shi, Osaka
c : Nomura Fudosan, Marubeni Corp.
e : Hironari Tanaka, Hiroki Nomaru, Osakazu Miyaji (Obayashi) / Yoshinori Yokoyama, Takashi Yoshida (Takenaka)
f : Obayashi Corp., Takenaka Corp.
g : SS Osaka
h : A traffic line plan which dissociate passers from automobiles and an exterior space establishing a mini-sanctuary intend human to commune with nature.
i : ILB pavement, paving stone tile, wash plane table block, field stone
j : Obayashi Corporation

a：コンフォート岡山
b：岡山県弓之町
c：コンフォート有楽
e：大成建設設計本部/小笠原祥仲
f：大成建設広島支店
g：村岡正巳（国際写真）
h：パブリックの環境を充実させる壁は，国道から視線と喧騒をさえぎり，ファサードに存在感を与える。
i：磁器質タイル，擬石吹き付け，黒御影石
j：大成建設設計本部
●

a：COMFORT OKAYAMA
b：Yumino-cho, Okayama
c：Comfort Yuraku Co.,Ltd.
e：Taisei Corporation Design and Proposal Division, Yoshinaka Ogasawara
f：Taisei Corporation Hiroshima Branch
g：Masami Muraoka(Kokusai Shashin Co.,Ltd.)
h：A wall which enriches public environment avoids public gaze and the din and bustle from the national road and makes a facade conspicuous.
i：Vitreous tile, sprayed-on imitation stone, black granite
j：Taisei Corporation Design and Proposal Division

a：URBAN VEGA
b：兵庫県神戸市
c：藤倉良行
d：池上俊郎
e：池上俊郎
f：コスモピア建設
g：長尾純之助
h：神戸市中心部幹線道路沿いの約60平方メートルの敷地である。塔状のメゾネットを中心とする集合住宅である。
i：無機塗装ALC板，無機塗装スチールパネル，ALC板アクソナイト塗装
j：池上俊郎＋アーバンガウス研究所
●

a：URBAN VEGA
b：Kobe-shi, Hyogo
c：Yoshiyuki Fujikura
d：Toshiroh Ikegami
e：Toshiroh Ikegami
f：Cosmopia Construction Company
g：Junnosuke Nagao
h：The site measures approximately 60 sqm and face the principal road of center of Kobe. It is an apartment house centering a tower-shaped maisonette.
i：Painting on ALC board
j：Toshiroh Ikegami+Urban Gauss

a：建築家が建てた建売住宅（朝霧住宅）
b：兵庫県明石市
c：マルコー
f：宗吉工務店
g：山田
h：建売住宅，なんと響きの悪い呼び名であろうか，そのなかに空間と町並を入れていく。
i：屋根：カラーベストコロニアル/外壁：モルタル刷毛引，リシン吹き付け
j：二谷重男＋U建築企画室
●

a：ASAGIRI NO.1
b：Akashi-shi, Hyogo
c：Maruko
f：Muneyoshi Office
g：Yamada
h：TATEURI JUTAKU (Ready-built house).How bad taste the name is！A group of house built for sale is designed to express the image of Specious and functional town.
i：Roof：colored bestos colonial.Exterior wall：Mor:ar brush finish pneumatic applied lysine
j：Shigeo Uetani+U Architectural Office

a：リバーテラス
b：大阪府大阪市
c：オーサカロイヤルアパートメント
e：竹中工務店設計部/吉田寛史
f：竹中工務店
g：松村芳治，古田雅文
h：低層部は，いちょう並木や堤防の緑と呼応するガレ
リアやポケットパークがつくられている。
i：外壁：45二丁磁器タイル，花こう岩 J&P/床：玉砂
利洗出，花こう岩 J&P
j：竹中工務店

●

a：RIVER TERRACE
b：Osaka-shi, Osaka
c：Osaka Royal Apartment Co.,Ltd.
e：Takenaka Corporation, Building Design
Dep./Hiroshi Yoshida
f：Takenaka Corporation
g：Yoshiharu Matsumura, Masafumi Koda
h：In the lower part of the building, the gallerie
and pocket park have been designed in
response to the gingko avenue and the
greenery of the bank.
i：Exterior walls : 45 double-ply ceramic tiles,
granite veneer J&P/Floor : scrubbed-finish
gravel stone, granite veneer J&P
j：Takenaka Corporation

a：エル・キャビン（長谷工コーポレーション新浦安独身寮）
b：千葉県浦安市
c：長谷工コーポレーション
d：中田文彦
e：中田文彦，永井幹子
f：長谷工コーポレーション
g：中田文彦
h：自社の独身寮のため，新入社員が本物を見極める
目を育てるための工夫を盛り込んだ設計とした。
i：外装：セラキューブ，インターロッキング／内装：ジ
ョリパット，マモリウム
j：長谷工コーポレーションエンジニアリング事業部

●

a：L CABIN
b：Urayasu-shi, Chiba
c：Haseko Corporation
d：Fumihiko Nakata
e：Fumihiko Nakata, Mikiko Nagai
f：Haseko Corporation
g：Funihiko Nakata
h：Being a bachelor dormitory, it is designed
to incorporate the idea a contrivance for new
employees to learn how to tell a genuine.
i：Exterior : ceramic cube, interlocking
j：Haseko Corporation Engineering Dept.

a：和多田邸
b：京都府城陽市
c：和多田光郎
f：久寿工務店
g：照屋
h：新興住宅地の建売住宅群のなかの風景を切り取ることに主眼をおく。
i：外壁／一部アルミ貼り
j：上谷重男＋U建築企画室
●
a：WATADA RESIDENCE
b：Joyo-shi, Kyoto
c：Mituro Watada
f：Kusu
g：Teruya
h：The principle matter is to cut off a house from the landscape sorrounded by a group of house built for sale.
i：Exterior wall : Partly aluminum foil
j：Shigeo Uetani+U Architectural Office

a：紅の家 NO.1
b：兵庫県神戸市
c：照屋涼子
f：久寿工務店
h：新しい住宅地に，子供たちのための風景を創りえたか。夕日に輝く紅の色のかがやき。
i：木造2階建
j：上谷重男＋U建築企画室
●
a：KURENAI NO.1
b：Koube-shi, Hyogo
c：Ryoko Teruya
f：Kusu Komuten
h：Is a landscape for children be able to created in a newly built residential area ?
KURENAI NO IE NO.1 (Deep red house No.1) is designed with an image of Glow of sunset.
i：Wood
j：Shigeo Uetani+U Architectural Office

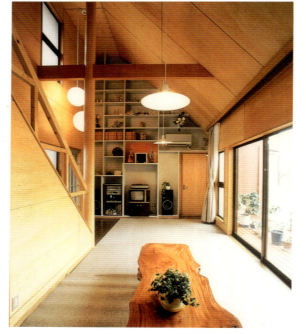

a：新日本製鉄，大和田社宅
b：千葉県君津市
c：新日本製鉄
e：渡辺康人（日本カラーテクノロジー研究所）
f：大栄
g：立見治嗣
h：大規模団地のリフレッシュ計画で，建物の配色を変えながら全体を個性的なイメージにまとめた。
i：アルミ
j：新日本製鉄君津製鉄所
●
a：NIPPON STEEL CORPORATION OWADA COMPANY HOUSE
b：Kimitsu-shi, Chiba
c：Nippon Steel Corporation
e：Yasuto Watanabe（Japan Color Technolgy Institute）
f：Daie
g：Harutsugu Tatsumi
h：A large-scale apartment development area is refreshed by changing the coloring of the buildings to integrate entire buildings' image to be characteristic as its refresh plan.
i：Aluminum
j：Nippon Steel Corporation Kimitsu Works

a：S邸
b：北海道札幌市
c：斉藤敬一
d：福井正和
e：福井正和
f：青木建設
g：細間邦明
h：札幌の夜景が見える高台に位置するロケーションで、全室より眺望を楽しめる設計。
i：タイル、石、曲面ガラス、ステンレス、ウレタン塗装、アルミ、シンチュウ
j：オリエント1級建築士事務所

●
a：S TEI
b：Sapporo-shi, Hokkaido
c：Keiichi Saito
d：Masachi Fukui
e：Masachi Fukui
f：Aoki Kensetsu
g：Kuniaki Hosoma
h：It is designed that each rooms has a fine view because it is located on an eminence, commanding a night view of Sapporo.
i：Tile, stone, curved surface glass, stainless, urethane painting, aluminum, brass
j：Orient Architects & Planners

a：堀口邸
b：大阪府吹田市
c：堀口正次郎
e：歌一洋，俵雅人
f：クレイデコ
g：松村芳浩
i：ガルバリウム鋼板，モルタル，コンクリート
j：歌一洋建築研究所
●
a：HORIGUCHI HOUSE
b：Suita-shi, Osaka
c：Syoujiro Horiguchi
e：Ichiyo Uta, Masato Tawara
f：LA KREIDECO
g：Yoshiharu Matsumura
i：Mortar, concrete
j：Ichiyo Uta Architect & Associates

a：ベルクルーエ小諸
b：長野県小諸市
c：阪口正一
d：菊地正典
e：菊地章
f：小林建設
g：松田博史
h：枯松と積雪が造り出す自然の調和よりむしろ各々
の存在する力強さに呼応させたものを設計した。
i：吹き付けタイル、エンコ板貼り
j：ディエム・デザイン
●

a：BELLECRUIE KOMORO
b：Komoro-shi, Nagano
c：Shoichi Sakaguchi
d：Masanori Kikuchi
e：Akira Kikuchi
f：Kobayashi Kensetsu
g：Hiroshi Matsuda
h：The vigorousness existing in each element
has been emphasized in the design of this
construction as opposed to utilizing the
harmony between natture created by the old
pine trees and snow.
i：Spray tiles, Enko board plaster
j：DM Design Co.,Ltd.

a：S邸
b：神奈川県横浜市
c：佐伯健一
d：菊地正典
e：菊地章
f：幸和建設
g：松田博史
h：光や風が住宅から失われていくなかで，高台に位
置するこの家に光が光である，風が風であることを願った。
i：ハーフボーダータイル
j：ディエム・デザイン
●
a：S HOUSE
b：Yokohama-shi, Kanagawa
c：Kenichi Saeki
d：Masanori Kikuchi
e：Akira Kikuchi
f：Kowa Kensetsu
g：Hiroshi Matsuda
h：Recent house are being built under the
conditions of less natural light and draught.
Conversely, this house was especially built on
high ground to ensure it gets a full amount of both.
i：Half border tiles
j：DM Design Co.,Ltd.

a：見富邸
b：静岡県熱海市
c：見富栄一
d：吉岡啓賀
e：吉岡啓賀
f：サン巧芸建設，鈴木孝志
g：吉岡啓賀
h：高低差8mの敷地条件下ながら極力自然を生かし，
自然と融合しつつも存在感のあるものを目指した。
i：伊豆自然石，間知石，御影飛石，サビ砂利，焼丸
太，竹，磁器タイル
j：サン吉岡建築設計事務所
●
a：MITOMI HOUSE
b：Atami-shi, Shizuoka
c：Eiichi Mitomi
d：Hiroshige Yoshioka
e：Hiroshige Yoshioka
f：San Kogei Kensetsu Co.,Ltd., Takashi
Suzuki
g：Hiroshige Yoshioka
h：A house is designed to exist harmonizing
with nature making full use of nature despite of
the project site condition which has 8m
difference.
i：Natural stone from Izu, granite stepping
stones, rust gravel, baked log, bamboo,
vitreous tile
j：San Yoshioka Architectural Office

a：大洋グループ軽井沢山荘
b：長野県北佐久郡軽井沢町
c：大洋興産
d：國分孝雄
e：國分孝雄
f：トーヨコ
g：三輪写真事務所
h：地域の景観と調和のとれた非日常的空間の演出
に留意した。
i：木造
j：国分設計 代表取締役國分孝雄
●
a：TAIYO GROUP KARUIZAWA SANSO
b：Karuizawa-cho, Kitasaku-Gun, Nagano
c：Taiyo Kosan Co.,Ltd.
d：Takao Kokubu
e：Takao Kokubu
f：Toyoko Co.,Ltd.
g：Niwa Shashin Jimusho
h：It is designed to represent unusual space
which harmonize with a view of the area.
i：Wood
j：Kokubu Sekkei Co.,Ltd. Architect TaKao
Kokubu

Chapter 6

サイン/ストリートファニチャー・デザイン
Sign, Street Furniture Design

サインのフォルムに近未来志向を見た
スペースワールドサイン・環境計画

株式会社丹青社サインデザイン研究所

北九州市八幡地区に出現したテーマパーク「スペースワールド」は、地球上のどこよりも宇宙に一番近い街、宇宙への出島（＝TANSIT POINT）をめざす近未来思考のアミューズメント施設である。宇宙への旅立ちに始まり、様々な冒険や宇宙的日常生活を体験できるテーマパークとして、4つのメインアトラクションと子ども向けのミニライド、スペースバザール、宇宙飛行士の訓練を体験できるスペースキャンプで構成されている。

サイン・環境計画では、景観設計のコンセプトである「宇宙の秩序との調和」を共通の理念としてデザイン開発が進められた。
そのコンセプトを具現化するために「丸」「三角」「四角」のようなプライマリーなかたちが施設全体に採用され、サインのフォルムにも構成要素として取り入れられている。さらに、次のようなエリアごとのイメージが付加されてそれぞれのエリアの特徴を表現している。

アプローチゾーン （Approach zone）

アプローチゾーン→パーキング→スペースシップ
パブリックゾーン→エントランス/スペースバザール/ゲストサービス→ステーション
アトラクションゾーン→アトラクション→プラネット
エデュケーションゾーン→スペースキャンプ→スペースラブ

そしてカラーリングについては，

　　　アーバンスケール（地上3メートル以上）でのシルバーメタリック。

　　　ヒューマンスケール（地上3メートル以下）でのペール＆カラフル。

　　　ワームスケール（地上0メートル）。

の三層からなる立面展開と，プリズム分光による色環を平面的に配置した
外部色彩計画に準じたカラー展開を実現した。またグラフィックデザインには
「秩序」あるデザインとシンプルなレイアウトを基本にし，
具体的にはストラップパターンで強調されたピクトグラムや，

長体で統一したサンセリフ書体を使用して適度な緊張感を持たせている。
エリアごとにフォルムやカラーリングが変化している中でベーシックな
グラフィックデザインを統一させて全体的な調和を計っている。
テーマパークのような特別に目的が限定されている施設のなかでの
サイン計画は，機能性に加えてそのテーマに即した演出効果や環境との調和が
要求される。また，個々の運営・サービスのコンセプトやマニュアルにまで
影響を及ぼすものであり綿密なサイン計画が
パーク全体の完成度を高める重要な要素となる。

Expectation to the near-future is found in the forms of sign.
Sign & Visual Environment of Space World

Tanseisha Sign Design Institute

The Space World theme park came into being in Yawata, Kita-Kyushu, and is an amusement facility based upon the concept of the near future in order to serve as a town that will be nearest to space and be able to become a transit point into space. It features four main attractions and is a theme park in which one can experience various space adventures and the type of daily life that goes on in space starting with the actual setting out on a space journey, enjoying a space camp in which children can go on the mini rides, wandering about a space bazaar and even experience the training necessary to become an astronaut.

During the sign and environmental plan, a design development was put forward which based itself on "harmony with the order of space"; a concept of the view design as a common idea. Primary shapes such as circles, triangles and squares were adopted in the facility in order to materialize this concept. These were utilized in the sign forms as componant elements. Additionally, each area was distinguished by the setting of the following images:

パブリックゾーン （Public zone)

Approach zone → Parking → Spaceshop
Public zone → Entrance/Space bazaar/Guest service → Station
Attraction zone → Attraction → Planets
Education zone → Space camp → Space lab

Also, the color design was based upon the external color plan that is arranged in a plain manner through the prismic spectrum. The graphic design was based on a simple and orderly layout; for example, the pictogram was emphasized by the use of a striped pattern, and long characters have been used to add a feeling of moderate tenseness. As each area is rich in variation of forms and coloring, basic graphic designs were used to maintain the entire balance.

Sign planning, when used in a facility such as a theme park for which the purpose is limited, requires a production effect based on a theme and must be in complete harmony with the environment as well as simply being functional. The given effect for individual management, service, concept and manuals requires minute sign planning in order to serve as an important element for the overall quality of the finished product.

アトラクションゾーン　(Attraction zone)

エデュケーションゾーン （Education zone）

作品名＝スペースワールドサイン・環境計画
所在地＝福岡県北九州市
クライアント＝株式会社スペースワールド
ディレクター＝斉藤明男（丹青社サインデザイン研究所）
デザイナー＝金子享子，岩崎芳夫（丹青社サインデザイン研究所），水口弘志，笹山賀司
施工者＝株式会社丹青社
撮影者＝田村康明，三富純，陣川信昭
応募代表者＝株式会社丹青社コーポレートコミュニケーションセンター

Project name=Sign & Visual Environment of Space World
Site=Kita-Kyushu-shi, Fukuoka
Client=Space World Co.,Ltd
Director=Akio Saito(Tanseisha Sign Design Institute)
Designer=Kyoko Kaneko, Yoshio Iwasaki (Tanseisha Sign Design Institute), Hiroshi Mizuguchi, Yoshiji Sasayama
Contractor=Tanseisha Co.,Ltd.
Photographer=Yasuaki Tamura, Jun Mitomi, Nobuaki Jinkawa
Applican=Tanseisha Corporate Communication Center

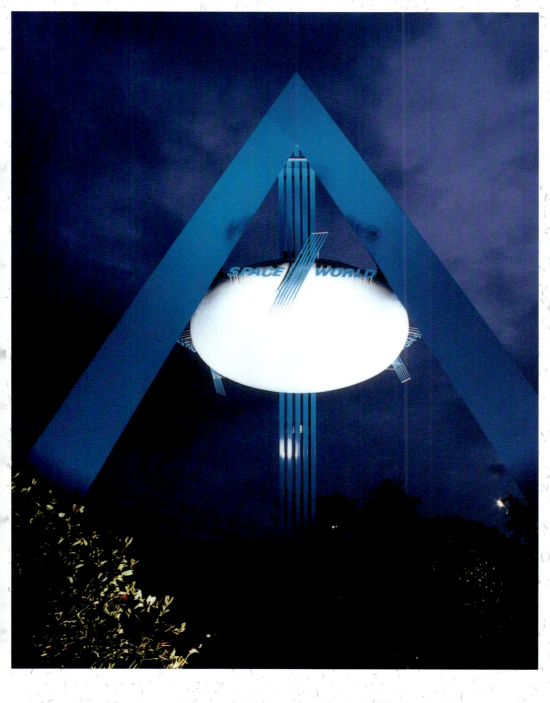

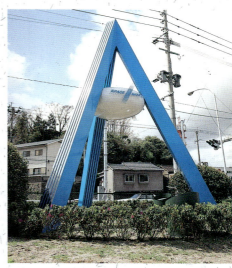

a：芝浦スクエアモニュメント「帆風」
b：東京都港区
c：竹中工務店
d：遠藤英雄
e：遠藤英雄，折笠智和
f：乃村工藝社
g：佐田山靖雄
h：芝浦スクエアビル。人間の体温を通さない硬質のインテリジェントビル。直線、直面、ガラス、ハイテク。この巨大な壁線に対して、一本の曲線のハイテンションの小宇宙が対峙する。直に対して曲、面に対して線、静に対して動、時間に対して悠久、風土の残像「帆」、ConpletionとTension。ガラスに映る芝浦の空の移り変わりが、風景の虚像と実像が、スクエアな都市空間の中で交歓する。
i：ステンレス
j：乃村工藝社
●

a : SHIBAURA SQUARE MONUMENT, FULL SAIL
b : Minato-ku, Tokyo
c : Takenaka Corporation
d : Hideo Endo
e : Hideo Endo, Tomokazu Orikasa
f : Nomura Co.,Ltd.
g : Yasuo Sadayama
h : Shibaura Square building.A solid intelligent building that dose not allow human temperature to pass through.This gigantic straight lineal, straight faced, glass and high-tech wall faces a curved high tensile smaller space.There are contrasts between the straight and cured, surface and lines, stillness and motion, time and eternity, the after images of natural features and sails and completion against tension.The variations of the Shibaura sky and the virtual and real images of the scenery reflected in the glass exchang greetings in this square city space.
i : Stainless steel
j : Nomura Co.,Ltd.

a：千葉市都市装置
b：千葉県千葉市
c：千葉市
d：石川晃二
e：金子清二，向隆宏，大友昭治
f：乃村工藝社
g：大東正巳
h：連続する5基の塔から姉妹友好都市を象徴する人形たちが登場する。塔上のカモメは人々を海へと誘う。
i：ステンレス
j：乃村工藝社
●

a : CHIBA CITY URBAN DEVICE
b : Chiba-shi, Chiba
c : Chiba City
d : Koji Ishikawa
e : Seiji Kaneko, Takahiro Mukai, Shoji Otomo
f : Nomura Co.,Ltd.
g : Masami Daito
h : Dolls symblizing the sister-city will appear from the connected 5 towers.The seals sitting on the top towers will lead people to the sea.
i : Stainless steel
j : Nomura Co.,Ltd.

a：朝日新聞100周年記念モニュメント
b：東京都中央区
c：竹中工務店
d：中辻伸
e：石井泰二
f：乃村工藝社
g：大東正巳
h：「時」を素材の転換により紡ぎあげ，文化の変遷を
語ると同時に，「時」の流れを封じ込めることを意図した。
i：自然石，御影石，ガラス，チタン
●
j：乃村工藝社

a：ASAHI SHINBUN CENTENNIAL MONUMENT
b：Chuo-ku, Tokyo
c：Takenaka Corporation
d：Shin Nakatsuji
e：Taiji Ishii
f：Nomura Co.,Ltd.
g：Masami Daito
h：Time has been spanned by the changing
materials, and at the same time as relating the
restoration of a culture, the passage oftime
has been designed to be confined within.
i：Natural stone, granite stone, glass, titanum
j：Nomura Co.,Ltd.

a：大山道再生計画一弦巻・池尻
b：東京都世田谷区
c：世田谷区
d：中辻伸
e：中辻伸，金子清二，宇田靖夫
f：乃村工藝社
g：佐田山靖雄
h：江戸庶民でにぎわった大山道。時を越えて，世田谷
の風土に根ざした体感コミュニケーションサイン。
j：乃村工藝社
●

a：OYAMA-MICHI SYMBOL SIGN
b：Setagaya-ku, Tokyo
c：Setagaya-ku
d：Shin Nakatsuji
e：Shin Nakatsuji, Seiji Kaneko, Yasuo Uda
f：Nomura Co.,Ltd.
g：Yasuo Sadayama
h：The Oyama-michi was once alive with
people in the Rdo era.More than 100 years
later it is a sensational body communication
sign rooted to the natural features of Setagaya.
j：Nomura Co.,Ltd.

a：足利市環境装置「歴史から未来へ」
b：栃木県足利市
c：足利市
d：中辻伸
e：石井泰二，向隆宏
f：乃村工藝社
g：大東正巳
h：渡良瀬川，両崖山に浅間山。足利一門の氏寺，ば
んな寺。足利の風土をモチーフに，新和風表現を構築。
i：黒御影石，発色ステンレス
j：乃村工藝社
●

a：ASHIKAGA MONUMENT
b：Ashikaga-shi, Tochigi
c：Ashikaga City
d：Shin Nakatsuji
e：Taiji Ishii, Takahiro Mukai
f：Nomura Co.,Ltd.
g：Masami Daito
h：The Watarase River, Mt.Ryogan, Mt.Asama,
the Ashikaga family temple.Using the natural
features of Ashikaga as a motif, a new style of
Japanese expression was constructed.
i：Black granite, color devloped stainless steel
j：Nomura Co.,Ltd.

a：神戸市メリケンパークファッションモニュメント
b：兵庫県神戸市
c：神戸市
d：奥田総一郎
e：奥田総一郎
f：木下工務店
h：神戸市の花アジサイの花の色を基調としたさまざ
まな色のモザイクで，華やかなイメージを演出した。
i：磁器質陶板
j：岩尾磁器工業
●

a：KOBE-SHI MERIKEN PARK FASHION
MONUMENT
b：Kobe-shi, Hyogo
c：Kobe-City
d：Soichiro Okuda
e：Soichiro Okuda
f：Kinoshita Komuten
h：Brilliant image is produced by various
colored mosaic based on the color of
hydrangea which is flowers of Kobe.
i：Vitreous tile
j：Iwao Jiki Kogyo Co.,Ltd.

a：日時計
b：茨城県土浦市
c：土浦市役所
e：大和測量
f：サカエ
g：井上和幸
i：本石，ステンレス
j：サカエ

a：SANDIAL
b：Tsuchiura-shi, Ibaraki
c：Tsuchiura City
e：Yamato Skuryo
f：Sakae Co.,Ltd.
g：Kazuyuki Inoue
i：Stones, Stainless steel
j：Sakae Co.,Ltd.

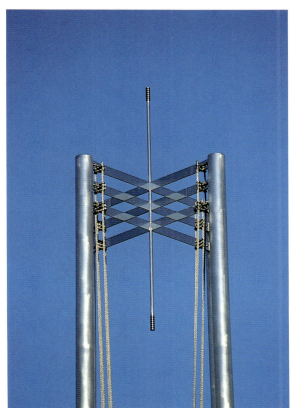

a：GRAVITY
b：大阪府大阪市
c：大阪大学理学部同窓会
d：池上俊郎
e：池上俊郎＋アーバンガウス研究所　担当：薮内秀美
f：リホミカ
g：池上俊郎
h：大阪大学理学部跡地記念碑。私たちの原点の一つである地球を重力で示すべく、宙空に浮く球を用意した。
i：土台：RC打っ放し小たたき仕上げ／照明器具埋込／円柱、ワイヤー、球、銘板：ステンレス
j：池上俊郎＋アーバンガウス研究所
●

a：GRAVITY
b：Osaka-shi, Osaka
c：Alumni Association of the Faculty of Science, Osaka University
d：Toshiroh Ikegami
e：Toshiroh Ikegami+Urban Guauss, Hidemi Yabuuchi
f：Rihomika
g：Toshiroh Ikegami
h：A monument for remains of Osaka University the science dapartment.A ball floating in the air is adopted to show the earth, one of our origin through expressing the gravity.
i：Ground sill : RC exposed hard-stone finish, recessed light fixture.A column, wire, a ball and a nameplate : Stainless steel
j：Toshiroh Ikegami+Urban Gauss

a：自然との対話
b：神奈川県平塚市
c：かながわ都市緑化平塚フェア実行委員会
e：スタジオ・ビー（原田昭久）
f：サカエ＋スタジオ・ビー
g：原田昭久
h：風によって可動部が変化し、形状記憶合金を使用した鈴が季節を感じ違った音色をだす。
i：ステンレス、本石、形状記憶合金
j：サカエ
●
a：COMMUNICATION WITH NATURE
b：Hiratsuka-shi, Kanagawa
c：Hiratsuka Fair Practice Committee
e：Studio Vee, Akihisa Harada
f：Sakae Co.,Ltd. and Studio Vee
g：Akihisa Harada
h：An adjustable part of a bell using shape memory compound metal is changed by wind makes different tone color by feeding seasons.
i：Stainless steel, stones, shape memory compound metal
j：Sakae Co.,Ltd.

a：西九州自動車道，波佐見・有田 IC バスストップの
　時計塔
b：長崎県波佐見町
c：日本道路公団福岡建設局
e：イワオ・ベセラ・デザインチーム
f：岩尾磁器工業
h：記念碑のエピソードや焼きものに関する説明などを
　表示し，観光客への案内の機能を持たせている。
i：磁器
j：岩尾磁器工業
●

a：NISHI-KYUSHU EXPRESS WAY HASAMI,
ARITA I.C.BUS STOP CLOCK TOWER
b：Hasami-machi, Nagasaki
c：Fukuoka Construction Bureau, Japan High
Way Public Corporation
e：Iwao Jiki Desigh Team
f：Iwao Jiki Kogyo Co.,Ltd.
h：A clock tower having functions of guidance
for tourists showing episodes of this
monument and explanations of pottery.
i：Vitreous
j：Iwao Jiki Kogyo Co.,Ltd.

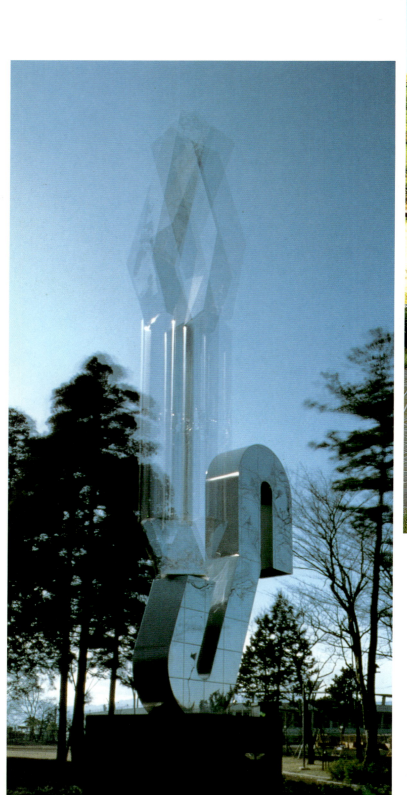

a：あつまろう緑の仲間たち
b：神奈川県相模原市
c：かながわ都市緑化相模原フェア実行委員会
e：スタジオ・ビー（原田昭久）
f：サカエ＋スタジオ・ビー
g：原田昭久
h：人工の動力を使わず風の流れによってモニュメント
自体が変化する。
i：ステンレス，本石，ポリカーボネイト，形状記憶合金
j：サカエ
●

a：ATSUMAROU MIDORI NO NAKAMATACHI
b：Sagamihara-shi, Kanagawa
c：Sagamihara Fair Practice Committee
e：Studio Vee, Akihisa Harada
f：SAKAE Co.,Ltd.+Studio vee
g：Akihisa Harada
h：A monument makes changes by the current
of the wind without any artificial power.
i：Stainless steel, stones, polycarbonate,
shape memory compound metal
j：Sakae Co.,Ltd.

a: イヌイビル・カチドキ・サイン計画
b: 東京都中央区
c: イヌイ建物
d: 竹中工務店
e: テイ・グラバー
f: テイ・グラバー
g: 輿水進
h: サインを空間構成の一環とし，建築外観をスケールダウンしたものをモチーフにデザイン。
i: ステンレス塗装仕上げ，アクリル
j: テイ・グラバー
●

a: SIGN PROJECT OF INUI BLDG.KACHIDOKI
b: Chuo-ku, Tokyo
c: Inui Tatemono
d: Takenaka-Corporation
e: T.Glover Co.,Ltd.
f: T.Glover Co.,Ltd.
g: Susumu Koshimizu
h: A sign is designed as a part of composing space using a motif which is scale downed exterior of the building.
i: Painted stainless, acryl
j: T.Glover Co.,Ltd.

a: 国際花と緑の博覧会 政府苑屋外サイン
b: 大阪府鶴見区
c: 国際花と緑の博覧会協会
d: 梶川伸二
e: 由井真波，富家大器，早川典臣
f: コトブキ
g: 斉藤さだむ
h: 修景の装置としてサインをとらえ，和の原風景と調和するような表現方法を試みた。
i: ステンレス，アルミ，アクリル，サツキ
j: GK京都
●

a: THE INTERNATIONAL GARDEN & GREENERY EXPO. JAPANESE GOVERNMENT PLAZA SIGN PROJECT
b: Tsurumi-ku, Osaka
c: Japan Association for the International Garden & Greenery Expo., Osaka, Japan, 1990
d: Shinji Kajikawa
e: Manami Yoshii, Taiki Tomiie, Noriomi Hayakawa
f: Kotobuki Seating
g: Sadamu Saito
h: An alternation of views has been attempted by fixing signs in orderto coordinate with typical view of Japanese fields.
i: Stainless steel, aluminium, acrylic, azalea
j: GK Kyoto Co.,Ltd.

a:国際花と緑の博覧会「会場全体サイン」
b:大阪府鶴見区
c:国際花と緑の博覧会協会
d:宮沢巧
e:梶川伸二、由井真波、大湊浩幸、早川典臣
f:コトブキ、日通商事、倉本産業
g:斉藤さだむ
h:花博のテーマである「自然と人間との共生」をアル
ミと樹木により表現。景観を秩序だてている。
i:ステンレス、アルミ、ホウキポプラ、ギンドロ
j:GK 京都
●
a:THE INTERNATIONAL GARDEN &
GREENERY EXPO. SIGN PROJECT
b:Tsurumi-ku, Osaka

c:Japan Association for the International
Garden & Greenery Expo., Osaka, Japan, 1990
d:Isao Miyazawa
e:Shinji Kajikawa, Manami Yoshii, Hiroyuki
Ominato, Noriomi Hayakawa
f:Kotobuki Seating, Nittsu Shoji, Kuramoto
Sangyo
g:Sadamu Saito
h:Aluminium and trees have been used to
express the theme of the Flower Exhibition,
MUTALISM BETWEEN NATURE AND PEOPLE.
The views are displayed in order.
i:Stainless steel, aluminium, broom populars,
Gindoro
j:GK Kyoto Co.,Ltd.

a:西宮観光30選説明サイン
b:兵庫県西宮市
c:西宮市役所企画局都市計画部都市計画課
d:宮沢功
e:森岡誠、平部喜弘
f:大昌工芸
g:山本圭一
h:時間の経過の中で素材の特性が表情として環境と
調和し、落ち着きを増すように造形は単純化した。
i:ステンレス、コールテン鋼、真鍮
j:GK 大阪
●
a:NISHINOMIYA, SELECTION OF 30 TOURIST
SPOTS-SIGN PROJECT
b:Nishinomiya-shi, Hyogo
c:Nishinomiya City, Urban Deveropment Office
d:Isao Miyazawa
e:Makoto Morioka, Yoshihiro Hirabe
f:Daisho Kogei Co.,Ltd.
g:Keiichi Yamamoto
h:The special quality of the materials have
been well-harmonized withthe surroundings as
face during the passing of time, and their
formative shapes have become simplified as
the amount of the scribbling increases.
i:Stainless steel, courderoy steel, brass
j:GK Osaka Co.,Ltd.

a:姫路城周辺地区総合サイン
b:兵庫県姫路市
c:姫路市姫路城周辺整備室
d:宮沢功
e:森岡誠、竹林香孝
f:コトブキ
g:山本圭一
h:「特別史跡姫路城跡」の指定地域の環境特性を十
分に考慮し、現代の材料で和風の造形を試みた。
i:ステンレス、チタン電解発色板
j:GK 大阪
●
a:HIMEJI CASTLE ZONE-TOTAL SIGN
PROJECT
b:Himeji-shi, Hyogo
c:Himeji City, Himeji Castle Zone
Improvement Office
d:Isao Miyazawa
e:Makoto Morioka, Yoshiyuki Takebayashi
f:Kotobuki Seating
g:Keiichi Yamamoto
h:As this area is regarded as a specially
designated area for the special historical site
of the Himeji Castle, the environmental features
have been fully considered. Consequently, the
models were made in the Japanese style, but
with the utilization of modern materials.
i:Stainless steel, titanium electrode
chromophoric sheeting
j:GK Osaka Co.,Ltd.

a：グリーンホール相模大野サイン計画
b：神奈川県相模原市
c：相模原市
d：日本設計
e：テイ・グラバー／エージェンシー：乃村工藝社
f：テイー・グラバー
g：輿水進
h：形や色のデザインエレメントを建築から抽出して使用し、建築との一体感と機能を両立した。
i：ステンレス、アクリル、アルミ、ガラス
j：テイ・グラバー
●

a：SIGN PROJECT OF GREEN HALL SAGAMIOHNO
b：Sagamihara-shi, Kanagawa
c：Sagamihara-City
d：Nihon Sekki
e：T.Glover, Nonura Co.,Ltd.
f：T.Glover Co.,Ltd.
g：Susumu Koshimizu
h：A sign is designed to consist with united feeling with the structure and the faculties by using design elements of forms and colors abstracting them from the structure.
i：Stainless, acryl, aluminum, glass
j：T.Glover Co.,Ltd.

a：千葉中央銀座ふれあい商店街モニュメント計画
b：千葉県千葉市
c：千葉中央銀座商店街振興組合
d：大成道路
e：テイ・グラバー
f：テイ・グラバー
g：輿水進
h：小さいはずの懐中時計をモニュメント化し訴求効果を高め、地域の話題づくりと活性化をねらった。
i：ステンレス、FRP、真ちゅう
j：テイ・グラバー
●

a：CHIBA CHUO GINZA SHOPPING STREET MONUMENT
b：Chiba-shi, Chiba
c：Chiba Chuo Ginza Shopping Street
d：Taisei-Doro
e：T.Glover Co.,Ltd.
f：T.Glover Co.,Ltd.
g：Susumu Koshimizu
h：Furnishing a topic and activation of the region by turning from a pocket watch usually supposed to be small into a monument to produce satisfactory results of sales promotion.
i：Stainless, FRP, Brass
j：T.Glover Co.,Ltd.

a：日東紡ビルサイン計画
b：東京都中央区
c：日東紡績
d：テイ・グラバー
e：テイ・グラバー
f：テイ・グラバー，建築設計：日本設計
g：輿水進
h：建築との一体化と機能性を重視し，可能な限り装飾的要素を排除した。
i：ステンレス、アルミ、アクリル
j：テイ・グラバー

a：SIGN PROJECT OF NITTOBO BLDG.
b：Chuo-ku, Tokyo
c：Nitto Boseki
d：T Glover Co.,Ltd.
e：T Glover Co.,Ltd.
f：T Glover, Nihon Sekkei
g：Susumu Koshimizu
h：A sign is designed to eliminate decoration factors as soon as possible with attaching importance to unite with the structure and to the faculties.
i：Stainless, aluminum, acryl
j：T Glover Co.,Ltd.

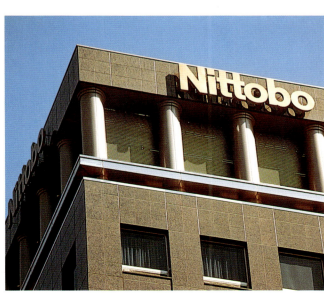

a：アーチ岩田
b：愛知県豊橋市
c：愛知商工
e：宮田正人
f：大幸建設名古屋支店
g：設計室アーチザン
h：このアーチ岩田は、東三河環状線沿いの東岩田町内に位置し、今回応募した虹のモニュメントは、環状線を通るドライバーたちのアイキャッチになっている。昨今のモータリゼーションに伴い増えつつある女性ドライバーが、親しみやすいサービスステーションのアイキャッチャーとしてモニュメントをデザインした。東三河環状線の中央分離帯のクスノキ並木からかかる虹のごとく、また湖西連邦にかかる虹のごとく、街道のランドマークとしてドライバーたちの目を楽しませている。
i：スチールパイプ、フッ素樹脂塗料吹き付け (New Gamet #5000)
j：設計室アーチザン
●
a：ARCH IWATA
b：Toyohashi-shi, Aichi
c：Aichishoko
e：Masato Miyata
f：Taikou Nagoya Branch
g：Artisan Architectural Design Office
h：Arch Iwata is located in a town called Higashi-iwata along the Higashi-mikawa loop highway.A Rainbow monument which is applied to this opportunity has been an eye-catcher for people passing through the loop highway.It is designed as a friendly eye-chatching monument of a service station for women drivers which nowadays are increasing in number due to motorization.The monument has been delighting driver's eyes as a landmark of the highway like the rainbow standing above the line of Kusu in the median strip of the Higashi-mikawa loop highway or like the rainbow appearing above the mountain range of Kosai.
i：Steel pipe, sprayed-on fluorine resin coating (New gamet # 5000)
j：Artisan Architectural Design Office

a：第一ホテルアネックスサイン計画
b：東京都千代田区
c：第一ホテル
d：テイ・グラバー
e：テイ・グラバー，建築設計：日本設計
f：テイ・グラバー
g：輿水進
h：ビジネス空間と調和しながら、都市生活のアメニティを演出。
i：アクリルシート、アルミヘアーライン、ウレタンちぢみ塗装
j：テイ・グラバー
●
a：SIGN PROJECT OF DAI-ICHI HOTEL ANNEX
b：Chiyoda-ku, Tokyo
c：Dai-ichi Hotel
d：T.Glover Co.,Ltd.
e：T.Glover Co.,Ltd., Nihon Sekkei
f：T.Glover
g：Susumu Koshimizu
h：A sign is designed to produce urban life amenity with harmonizing to a business space.
i：Acryl sheet, aluminum hair line, urethane, shrinkage painting
j：T.Glover Co.,Ltd.

a：スパリゾートハワイアンズ
b：福島県いわき市
c：常磐興産
d：テイ・グラバー
e：テイ・グラバー，日本設計 (建築設計)
f：テイ・グラバー
g：輿水進，安部順
h：地域に根ざしたアイデンティティの確立を目指し、サインからステーショナリーまでトータライズを行った。
i：ステンレス塗装仕上げ，アクリル塗装仕上げ，塩ビシート
j：テイ・グラバー
●
a：SPA RESORT HAWAIIANS
b：Iwaki-shi, Fukushima
c：Jobankosan
d：T.Glover Co.,Ltd.
e：T.Glover Co.,Ltd., Nihon Sekkei
f：T.Glover Co.,Ltd.
g：Susumu Koshimizu, Jun Abe
h：From its sign to its stationary are coordinated to establish and identity taking a deep hold upon the region.
i：Stainless painting finish, acryl painting finish, polyvinyl chloride sheet
j：T.Glover Co.,Ltd.

a：ユーノス・サイン・システム
b：全国各地
c：マツダ
d：秋山茂樹
e：原尚正，高山稔，高野菊男
f：乃村工藝社
g：Work Shop／深見耕一
h：最上級車種を扱うディーラーとしての高級感・高品質感、そしてヨーロッパ感覚の伝わる点に努力した。
i：鉄，ステンレス，アルミ，パナグラフィックス
j：乃村工藝社
●
a：EUNOS SIGN SYSTEM
b：The whole country
c：Mazda Motor Corporation
d：Shigeki Akiyama
e：Naomasa Hara, Minoru Takayama, Kikuo Takano
f：Nomura Co.,Ltd.
g：Work Shop／Koichi Fukami
h：Much effort was expended to create a high class impression of a car dealer handling only the most expensive cars with an air of European good quality.
i：Iron, stainless steel, aluminum, panagra fix
j：Nomura Co.,Ltd.

a：グランシティ大津ヶ丘
b：千葉県沼南町
c：住宅・都市整備公団　東京支社
d：村岡賢二
e：小野里宏，貝山秀明，近代造形，モリチュウ
f：東光園緑化
g：住宅・都市整備公団／小野里宏
h：「都市，自然そしてゆとり」を基本テーマとした集合住宅。
i：耐候性鋼／ステンレス／光ファイバー／銅
j：アーク造園設計事務所
●
a：GRAND CITY OTSUGAOKA
b：Shonan-machi, Chiba
c：The Housing and Urban Development Corporation Tokyo Branch Office
d：Kenji Muraoka
e：Hiroshi Onozato, Hideaki Kaiyama, Kindai Zoukei Inc., Morichu Inc.
f：Tokoen Ryokuka Inc.
g：TheHousing and Urban Development Corporation/Hiroshi Onozato
h：A collective housing which is built based on the theme Urban space, nature and leeway.
i：Stainless steel, optical fibers, copper
j：LRC Landscape Design Office

a：大野城市「まどか・ふれあい通り」擁壁レリーフ
b：福岡県大野城市
c：大野城市
d：岩尾エンジニヤリング
e：イワオ・特販デザインチーム
f：岩尾磁器工業
h：児童による絵付白磁陶板約2,600枚と立体感あるレリーフで明るく楽しいイメージを表現した。
j：岩尾磁器工業
●

a：ONOJYO-SHI, MADOKA FUREAI DRI RELIEF
b：Onojyo-shi, Fukuoka
c：Onojyo-City
d：Iwao Engineerig Co.,Ltd.
e：Iwao Jiki Tokuhan Design Team
f：Iwao Jiki Kogyo Co.,Ltd.
h：A bright and pleasant image is represent by 2,600 sheets of white vitreous tiles painted by children and a three-dimensional relief.
j：Iwao Jiki Kogyo Co.,Ltd.

a：広島市観光サイン
b：広島県広島市
c：広島市
e：GK設計
f：大昌工芸
g：渡辺敏正
h：景観上好ましい都市空間を創造し、市民、外来者にとって快適で安心して歩ける誘導サインを考察。
i：アルミ押出型材、ステンレス、アルミ鋳物、御影石
j：大昌工芸
●

a：HIROSHIMA SIGN FOR SIGHTSEEING
b：Hiroshima-shi, Hiroshima
c：Hiroshima-City
e：GK Sekkei Associates
f：Taisho Kogei Co.,Ltd.
g：Toshimasa Watanabe
h：An induction sign is designed with consideration to create a pleasant urban space and be able to have a comfortable and secure walk for the citizens and visitors.
i：Aluminum extrusion materials, stainless steel, aluminum casting, granite
j：Taisho Kogei Co.,Ltd.

a：牛久自然観察の森
b：茨城県牛久市
c：牛久市役所産業部みどり課
d：学習研究社 環境企画設計室，猪狩晃一
e：猪狩晃一，石井敏和
f：スペースデザイン ハッピー
g：石井敏和
i：杉丸太，鉄骨，ステンレス，シンチュウ，カッティングシート，アルフォト板
j：スペースデザイン ハッピー
●

a：USHIKU NATURE SANCTUARY
b：Ushiku-shi, Ibaragi
c：Ushiku City Office
d：Gakken, Koichi Igari
e：Koichi Igari, Toshikazu Ishii
f：Space Design Happy
g：Toshikazu Ishii
i：Japan ceder log, structural steel, stainless steel, brass, cutting sheets
j：Space Degin Happy

a：マキシーショッピングセンター・リニューアル
b：沖縄県那覇市
c：合資会社宝観光開発
e：歌一洋，平野正治，俵雅人，櫛谷諒
f：佐川建設，佐藤工業
g：歌一洋
i：スチール，ステンレス，アルミ
j：歌一洋建築研究所
●

a：MAXY SHOPPING CENTER RENEWAL
b：Naha-shi, Okinawa
c：Takara Kanko Kaihatsu Co.,Ltd.
e：Ichiyo Uta, Masaharu Hirano, Masati Tawara, Akira Kushitani
f：Sagawa Corp., Sato Kogyo Construction Co.
g：Ichiyo Uta
i：Steel, stainless steel, aluminum
j：Ichiyo Uta Architect & Associates

a：喫煙テーブル
d：神部千太郎
e：中村雅人，片山賢
f：ダイチ
h：某研究所リフレッシュゾーンにおける，移動可能煙清浄機付スモーキングテーブル。
i：スチールパイプ，テントクロス，ハロゲンランプ，空気清浄機
j：日建設計
●

a：SMOKING TABLE
d：Sentaro Kanbe
e：Masato Nakamura, Ken Katayama
f：DAICHI
h：This is a smoking table with a built-in moblile smoke extractor located in a refreshing area of the research institute.
i：Steel piping, tent cloth, halogen lamps, air cleaner
j：Nikken Sekkei Ltd.

a：ロッテワールド・サイン計画
b：大韓民国ソウル特別市
c：ホテルロッテ，ロッテショッピング
d：乃村工藝社/大熊俊隆，芝田良治
e：乃村工藝社/織田誠爾，横塚潔，竹尾健一，土屋
勝彦，大橋成実，川村清志，西山千明，佐藤真由
美，小城由加子，LOTTE 蚕室建設本部
f：大地，第一 NEON 社，RAIMBOW AD.&
INTERIOR CO.益山，三和広告，光仁，
YOUNGHAN TOTAL Co.,Ltd.，コートク
g：川本斉
h：来場者に対し，各施設とその構成要素の成り立ちと
位置がわかりやすいように心掛けた。
j：乃村工藝社
●

a：LOTTE WORLD SIGNAGE PLANNING
b：Seoul, Korea
c：Hotel Lotte Co.,Ltd., Lotte Shopping Co.,Ltd.
d：Nomura Co.,Ltd./Toshitaka Okuma, Ryoji
Shibata
e：Nomura Co.,Ltd./Seiji Orita, Kiyoshi
Yokozuka, Kenichi Takeo, Katsuhiko Tsuchiya,
Shigemi Ohashi, Kiyoshi Kawamura, Chiaki
Nishiyama, Mayumi Sato, Yukako Kojo, Hotel
Lotte Co.,Ltd., Jamsil Project
f：Daichi Co.,Ltd., Cheil AD.Neon Sign Co.,
Ltd., Raimbow AD.& Interior Co.,Ltd., Iksan Co.,
Ltd., Sanwa Kokoku Co.,Ltd., Kojin Co.,Ltd.,
Younghan Total Co.,Ltd., Kohtoku Co.,Ltd.
g：Hitoshi Kawamoto
h：We have strived to construct a building in
which visitors can easily find the facilities or
composite elements in which they are interested.
j：Nomura Co.,Ltd.

a：大妻女子大学
b：東京都千代田区
c：学校法人大妻学院
d：吉川昭
e：中村雅人
f：Seiko, 高島屋
h：校舎3棟に囲まれたアトリウムに、学生達の憩いの場とするためシンボリックな時計塔を計画した。
i：床/フローリング，壁/一部大理石，時計塔/スチールパイプにメタリック塗装，木部/化粧練付
j：日建設計

a : OHTSUMA WOMEN'S UNIVERSITY
b : Chiyoda-ku, Tokyo
c : Ohtsuma Gakuin
d : Akira Yoshikawa
e : Msato Nakamura
f : Seiko, Takashimaya
h : A symbolic clock tower has been designed in the atrium of which three sides are surrounded by the school buildings in order to create a place of recreation and relazation.
i : Floor/flooring, wall/marble in part, clock tower/metalic coating on steel pipe, wooden parts/decorated plaster
j : Nikken Sekkei Ltd.

a：秋田総合生活会館・美術館「アトリオン」
b：秋田県秋田市
c：日本生命保険相互会社，秋田県，秋田市
d：茅野秀真
e：オブジェ・デザイン/ヴィンチェンツォ・イアヴィコリ，
マリア・ルイザ・ロッシ
f：フジタ工業他 J.V.，ナスエンジニアリング
g：SS 東北，ジー・バイ・ケー
h：冬の長い北国のアトリウムに，空間的な変化と広が
りをあたえ楽しくにぎわいのある環境を演出。
i：アルミ，ステンレス，ポリカーボネート板
j：日建設計
●

a : AKITA CULTURAL HALL & ART MUSEUM,
ATRION
b : Akita-shi, Akita
c : Nippon Life Insurance Company, Akita
Prefecture, Akita City
d : Hozuma Chino
e : Object/Vincenzo Iavicoli, Maria Luisa Rossi
f : Fujita Co., & Others J.V., Nas Engineering Co.,Ltd.
g : SS Tohoku, G BY K Co.,Ltd.
h : A place with spacious variations and
bustling happiness has been created as a
Cultureal Hall and Art Museum in the snow
country that experiences long winters.
i : Aluminium, stainless steel, polycarbonate boards
j : Nikken Sekkei Ltd.

234

Chapter 7

その他
Others

a：大本山總持寺三松閣
b：神奈川県横浜市
c：大本山總持寺
d：東京大学/稲垣栄三，岸谷孝一，鈴木成文
e：究建築事務所/千本木隆芳
f：大成建設
g：三輪晃久写真研究所
h：禅宗庫院建築の古規に従った切妻様式。現代建築の合理性，機能性を満たした精神的宗教空間。
i：銅板，フッソ樹脂，イタリア産御影石，尾州・台湾 桧集成材，着色岩綿吸音板
j：究建築事務所

●

a：DAIHONZAN-SOJI-JI SANSHOUKAKU
b：Yokohama-shi, kanagawa
c：Daihonzan-Soji-ji
d：Tokyo University/Eizo Inagaki, Koichi Kishitani, Shigefumi Suzuki
e：Kyu Architectural/Design Office, Takayoshi Senbongi
f：Taisei Corporation
g：Kokyu Miwa Atchitectural Photograph Laboratory
h：Gable style observing traditional regulations
of the Zen sect kuriin style architecture.
Spiritual religion space filling rationality and function of modern architecture.
i：Copper plate, fluorine resin, Italian granite,
Bishu and Taiwan cypress glue-laminated
timber, coloration rock wool sound absorption board
j：Kyu Architectural Design Office

a：ISS X 東光寺本堂新築工事
b：兵庫県尼崎市
c：東光寺
d：石田信行
e：石田信行
f：月森工務店
g：石田信行
h：伝統様式現代様式競合混合摩可不思議。南無阿弥陀仏。
i：屋根：いぶし瓦/壁：打っ放しコンクリート一部アクリルリジジン吹き付け/内部壁：吹き付け聚楽
j：ISS建築設計事務所

●

a：ISS X TOKO-JI
b：Amagasaki-shi, Hyogo
c：Toko-ji
d：Nobuyuki Ishida
e：Nobuyuki Ishida
f：Tsukimori Co.,Ltd.
g：Mobuyuki Ishida
h：It is designed to express a mysterious atmosphere mixing and competing a traditional style with a modern style.
i：Roof：Oxidized tile, Wall：Exposed concrete partly pneumatically applied acryl lysine, Interior wall：Spray-on Juraku
j：ISS Architects & Engineers Associates

a：嶽之下宮御社殿
b：静岡県駿東郡
c：嶽之下宮
e：竹中工務店設計部／三井富雄，嶋俊己
f：竹中工務店，大成建設
g：アーキフォトKATO／加藤敏明
h：金色に輝く大屋根の下，カーテンウォールを通して，神，人，自然が向きあう現代の神社建築。
i：大屋根：チタン板電解発色，アルミカーテンウォール／扇柱：結晶ガラス
j：竹中工務店
●

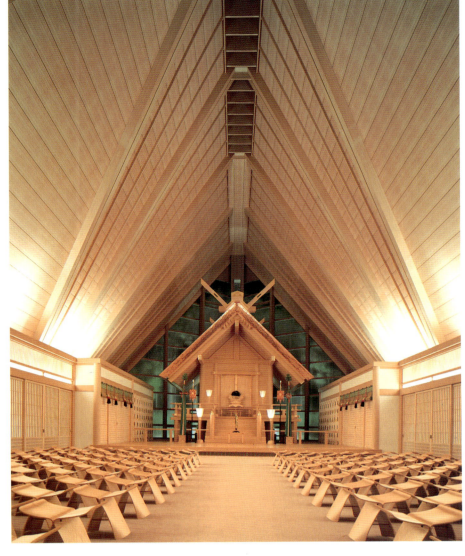

a : TAKENOSHITA SHRINE
b : Sunto-gun, Shizuoka
c : Takenoshita Religious Organization
e : Takenaka Corporation, Building Design Dep./Tomio Mitsui, Toshiki Shima
f : Takenaka Corporation, Taisei Corporation
g : Archi Photo Kato/Toshiaki Kato
h : A modern shrine in which God, the people and nature will confront one another through the curtain wall beneath the dome which glows with a golden color.
i : Dome : Electrolytic colored plate, aluminium curtain wall/Pillar : Crystal glass
j : Takenaka Corporation

a：西天神社新築工事
b：兵庫県伊丹市
c：西天神社護持会
f：絹川工務店
g：尾崎保雄
h：日本の原風景を創ってきた，鎮守の社。それがもっている空間の神秘性を現代の素材で。
i：鉄筋コンクリート平屋建
j：上谷重男＋U建築企画室
●

a : NISHITEN-JINJA
b : Itami-shi, Hyogo
c : Nishiten-Jinja Gojikai
f : Kinugawa Office
g : Yasuo Ozaki
h : A shinto shrine which has been creating Japanese primitive landscape.The spaceof its mysterious atmosphere is represented by using modern materials.
i : Reinforced concrete construction
j : Shigeo Uetani+U Architectural Office

a：法隆寺金堂壁画模写展示館
b：愛知県長久手町
c：法隆寺金堂壁画模写展示館建設協議会
e：竹中工務店設計部/吉岡一郎，古田智二，宮武仁
f：竹中工務店
g：車田写真事務所/車田保
h：古典のイメージを現代の手法で再構成し，近代的なキャンパスのなかに挿入した。
i：八分二丁磁器タイル，花こう岩，ステンレス，アルミキャスト
j：竹中工務店
●

a : HORYU-JI KONDO, REPRODUCTION OF FRESCOES MUSEUM
b : Nagakute-cho, Aichi
c : Horyu-ji Kondo, Reproduction of Frescoes Museum Building Conference
e : Takenaka Corporation, Building Design Dep./Ichiro Yoshioka, Toshiji Furuta, Jin Miyatake
f : Takenaka Corporation
g : Kurumada Photograph Office/Tamotsu Kurumada
h : The classic image of the building was reconstructed in modern methods and placed in the campus.
i : Double-ply ceramic tiles, granite stone, stainless steel, aluminium cast
j : Takenaka Corporation

a：神奈川トヨタ本牧ショールーム　カレリア
b：神奈川県横浜市
c：神奈川トヨタ
d：電通
e：乃村工藝社，川勝英史，谷田訓子
f：乃村工藝社
g：大東正巳
h：コミュニケーションの場のショールームを意図し，カフェ・カーショップ，多目的スペースを併設。
i：照明ポール/GRC吹き付け塗装，ミラーパーティション，什器天板/木象眼
j：乃村工藝社
●

a：KANAGAWA TOYOTA SHOWROOM CARELIA
b：Yokohama-shi, Kanagawa
c：Kanagawa Toyota Motor Corporation
d：Dentsu Inc.
e：Nomura Co.,Ltd., Hidefumi Kawakatsu, Kuniko Tanida
f：Nomura Co.,Ltd.
g：Masami Daito
h：This communication showroom was designed with the combined creation of a cafe, a car shop and a multi-purpose space.
i：Lighting pole/GRC spray paint, mirror partitions, fixture ceiling board/wooden inlaid work
j：Nomura Co.,Ltd.

a：トヨタ自動車池袋ビル AMLUX
b：東京都豊島区
c：トヨタ自動車
d：木谷靖孫
e：海宝幸一，高野勝也
f：竹中工務店
g：和木通，マツダ・プロカラー
h：情報伝達の一手段として採用されたコンピュータ
個別制御による，新型外壁照明システム。
i：アルミ押出し型材，無電極ランプ
j：日建設計
●

a : TOYOTA MOTOR IKEBUKURO BUILDING, AMLUX
b : Toshima-ku, Tokyo
c : Toyota Motor Corporation
d : Yasuhiko Kitani
e : Koichi Kaiho, Katsuya Takano
f : Takenaka Corporation
g : Toru Waki, Matsuda Pro-Color
h : A new exterior wall lighting system that is
controlled by an individual computer that has
been adapted as a method of communication.
i : Aluminium push-out material, non-electrode lamps
j : Nikken Sekkei Ltd.

a：トヨタ・オートサロン アムラックス東京
b：東京都豊島区
c：トヨタ自動車，電通
d：エス・エフ・メーカーズ
e：川勝英史，福田吉宏
f：乃村工藝社
g：大東正巳
h：トヨタの現在から未来が見えるショールーム。5フロ
アーの中に約70台の展示車とアミューズメント。
i：スチール焼付け塗装，ボンスエード塗装，フェロッ
クストーン仕上げ，ステンレス鏡面仕上げ
j：乃村工藝社
●
a：TOYOTA AUTOSALON AMLUX TOKYO
b：Toshima-ku, Tokyo
c：Toyota Motor Corporation, Dentsu Inc.
d：SF Makers
e：Hidefumi Kawakatsu, Yoshihiro Fukuda
f：Nomura Co.,Ltd.
g：Masami Daito
h：A showroom displaying the present into the
future of Toyota.Within the five doors there is
space to accomdate 70 display cars and amusements.
i：Steel sintering, bon suede paint, ferock
stone finish, stainless mirror surface finish
j：NOMURA Co.,Ltd.

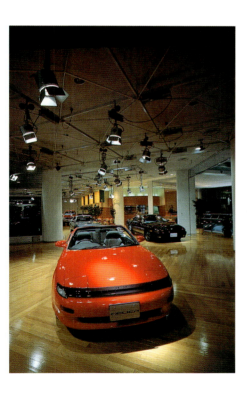

a：日産プリンス京都本社ショールーム
b：京都府京都市
c：日産プリンス京都販売
d：神尾勝宏
e：中村勲
f：船場
g：福本正明（アトリエ フクモト）
h：ショールームが乱立するこの地域で，特にファサード，内部空間に他社との明確な差別化を行った。
i：グレーチング敷黒目地仕上げ，鉄板8mm 焼付テーブルトップ，モルタル壁現場指示仕様，他
j：船場
●
a：NISSAN PRINCE KYOTO HEAD OFFICE
SHOWROOM
b：Kyoto-shi, Kyoto
c：Nissan Prince Kyoto Sales Co.,Ltd.
d：Katsuhiro Kamio
e：Isao Nakamura
f：Semba Corporation
g：Masaaki Fukumoto（Atelier Fukumoto）
h：A showroom is designed to make an
obvious distinction from others in the area
which flooded with showrooms especially
facade and interior space.
i：Laying grating black pointing finish, steel
plate 8m/m baked tabletop, mortar wall, others
j：Semba Corporation Design Section

a：日産アプリーテ千葉
b：千葉県千葉市
e：竹中工務店設計部/久林正晴，倉谷寛，広島正明
f：竹中工務店
g：ミヤガワ/藤田幹夫
h：周辺のコンテクストに対し，aprite なイメージとアイデンチティを表現した。
i：外壁：ALC，アルミ，スチール/内装：スチールメッシュ，松，ホモジニアスタイル
j：竹中工務店
●
a：NISSAN APRITE CHIBA
b：Chiba-shi, Chiba
e：Takenaka Corporation, Building Design
Dep./Masaharu Hisabayashi, Hiroshi Kuratani,
Masaaki Hiroshima
f：Takenaka Corporation
g：Miyagawa Photo/Mikio Fujita
h：Blending with the surroundings, this
construction has been built to express a pretty
image and identity.
i：External wall：ALC, aluminium, steel/
Interior：steel mesh, pine, homognia style.
j：Takenaka Corporation

a：アトラス長門
b：山口県長門市
c：丸久
d：武井淳
e：桐岡栄，大下稔夫，田中政明，ロータリーコーポレーション，中央設計
f：前田建設工業・住友建設共同企業体
g：カルダン・プロ
h：地域密着の生活広場として，施設を位置付け，駐車場も1,000台収容可能である。
i：カラーガルバリウム鋼板
j：乃村工藝社
●

a：ATRAS NAGATO
b：Nagato-shi, Yamaguchi
c：Marukyu Co.
d：Atsushi Takei
e：Sakae Kirioka, Toshio Oshita, Masaaki Tanaka, Rotary Corporation, Chuo Sekkei Co.,Ltd.
f：Maeda Kensetsu Co.,Ltd.& Sumitomo Kensetsu Co.,Ltd., J.V.
g：Cardin Pro Co.,Ltd.
h：As this is an indespensible square for local livelihood, a car park facility with a 1,000 car cpacity has been created.
i：Color galvanized steel board
j：Nomura Co.,Ltd.

a：もめんや藍
b：三重県伊勢市
c：赤福
e：竹中工務店設計部/植田明，宮野裕，岩下啓和
f：竹中工務店
g：青山写真館
h：「松坂もめん」ブティックショップを妻入り，きざみ囲いの伝統建築技法による町家で街並みに融合させた。
i：ツガ材，杉材，砂壁，花こう岩，日本瓦，和紙，三和土
j：竹中工務店
●

a：MOMENYA AI
b：Ise-shi, Mie
c：Akafuku
e：Takenaka Corporation, Building Design Dep./Akira Ueda, Yutaka Miyano, Yoshikazu Iwashita
f：Takenaka Corporation
g：Aoyama Shashinkan
h：The boutique MATSUZAKA COTTON was arranged in a traditional tradesman's house style of architecture in order to intergrate it with the row of stores and house along the street.
i：Plum tree timber, pine tree timber, sandcoat walls, granite, Japanese slate, Japanese rice paper, Miwa-tsuchi
j：Takenaka Corporation

a：佐世保四ヶ町商店街アーケードの方位盤
b：長崎県佐世保市
c：佐世保市
d：船場，佐世保市，計量計画研究所
e：イワオ・特販デザインチーム
f：松尾建設
h：タイムカプセル埋設地に佐世保の象徴カモメを中心とした全面フリーモザイクで方位盤を設置した。
i：磁器
j：岩尾磁器工業
●

a：SASEBO, YONKA-CHO, ARCADE COMPASS
b：Sasebo-shi, Nagasaki
c：Sasebo-City
d：SEMBA, Sasebo-shi, The Insutitute of Behavioral Sciences
e：Iwao Jiki Tokuhan Design Team
f：Matsuo Constraction Company
h：An azimuth compass of which surface is made of mosaic with a sea gull which is symbol of Sasebo at the central figure at a place where time sapsules are burried.
i：Ceramics
j：Iwao Jiki Kogyo Co.,Ltd.

a：寒霞渓パブリックトイレ
b：香川県小豆郡
c：小豆島総合開発
d：福家克彦
e：戸塚元雄，長本朝子
f：佐伯建設
g：木下保
h：標高600mの崖が切り立った高地の観光地で，明解な幾何学形態とすることにより厳しい自然に対峙している。
i：補強レンガブロック，御影石，ビニールタイル，ガラスブロック
j：慧匠社建築研究所
●
a：KANKAKEI PUBLIC TOILET
b：Shodo-gun, Kagawa
c：Shodoshima Sogo Kaihatsu Co.
d：Katsuhiko Fuke
e：Motoo Tozuka, Asako Nagamoto
f：Saeki
g：Tamotsu Kinoshita
h：A public lavatory, built in a high land tourist spot having a 600 meter sheer cliff above the sea, takes a stand against severe nature by forming a geometrical figure clearly.
i：Reinforced brick block, granite, vinyl tile, glass block
j：Keishosha Archi-Tect

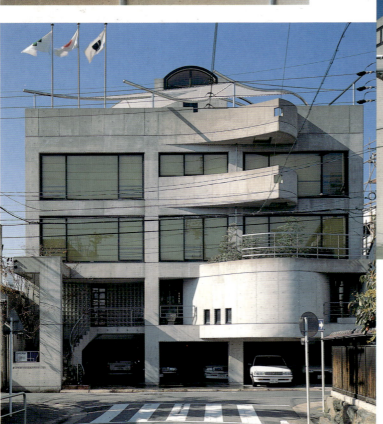

a：未完成ビル屋上テント増築
b：愛知県名古屋市
c：加納工務店
d：加納隆（創和コーポレーション）
e：加藤園子（加納建築設計事務所）
f：加納総建サービス
g：牧野裕（そあスタジオ）
h：「母」をテーマとするビルに会社創立40周年を記念して，「髪」をモチーフに新風を呼ぶテントをデザイン。
i：ステンレスヘアライン仕上げ，テント
j：加納建築設計事務所
●

a：UNFINISHED BUILDING'S TENT
b：Nagoya-shi, Aichi
c：Kanoh Corp.
d：Takashi Kanoh (Sowa Corporation)
e：Sonoko katoh (Kanoh Architectural Office)
f：Kanoh Soken Service
g：Yutaka Makino (Soar Studio)
h：A tent which call in a new phase using a motif TRESSES is designed on a building which theme is MOTHER for the 40th anniversary of the founding of the company.
i：Stainless hair line finish, tent
j：Kanoh Architectural Office

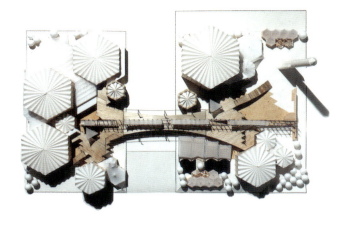

a：横浜博覧会テーマゾーン
b：神奈川県横浜市
c：横浜博覧会協会
d：石見修一，杉本洋文
e：竹中工務店設計部/岩崎堅一，田島綾夫，田中孝
次，橋本博，永島亮太郎，荻原剛，坂田克哉
f：竹中工務店
g：堀内広治
h：宇宙と子供たちをテーマに，視覚的変化，リズム感
のある空間の創出を意図した。
i：屋根：PUCコーティングポリエステル繊維布/外
壁：プライウッドサイディング
j：竹中工務店
●

a : YOKOHAMA EXOTIC SHOWCASE '89,
THEME ZONE
b : Yokohama-shi, Kanagawa
c : Yokohama Exotic Showcase Comunity
d : Shuichi Ishimi, Hirofumi Sugimoto
e : Takenaka Corporation, Building Design
Dep./Kenichi Iwasaki, Yasuo Tajima, Takatsugu
Tanaka, Hiroshi Hashimoto, Ryotaro
Nagashima, Takeshi Hagiwara, Katsuya Sakata
f : Takenaka Corporation
g : Koji Horiuchi
h : With A Spaec and Children as a theme, the
creation of a well-coordiated space rich in
visual variations and rhythms was the target.
i : Roof : PUC coated polyestter fiber cloth/
Exterior wall : Plastic wood siding
j : Takenaka Corporation

a：大阪市咲くやこの花館
b：大阪府大阪市
c：大阪市
d：日建設計/橋本巖
e：日建設計/塩井保則
f：竹中工務店
g：共同プロ（航空写真）
h：熱帯から高山植物まで8つのゾーンからなる温室内
の植栽は、外部庭園へと連続し世界の環境を体験できる。
i：アルミサッシ、ガラス
j：日建設計
●

a : GREAT CONSERVATORY, SAKUYA
KONOHANA KAN
b : Osaka-shi, Osaka
c : Osaka City
d : Nikken Sekkei/Iwao Hashimoto
e : Nikken Sekkei/Yasunori Shioi
f : Takenaka Corporation
g : Kyodo Pro
h : The plants within the greenhouse consist of
8 different zones from the tropics to alpine
areas and continue to the outside garden
which enables one to experience the different

enviroments of teh world.
i : Aluminium sashes, glass
j : Nikken Sekkei Ltd.

a：国際花と緑の博覧会 政府苑「都市・環境」館
b：大阪府大阪市
c：電通，東急エージェンシー J.V.
d：濱田隆法
e：濱田隆法，上田万寿，長沢信夫
f：乃村工藝社
g：大東正巳
h：現代都市におけるさまざまな緑化と，花・緑の重要な役割や生き物の共生をテーマにしている。
i：FRP成形，アクリル成形，木工造形，可動装置
j：乃村工藝社
●

a : THE INTERNATIONAL GARDEN AND GREENERY EXPOSITION, JAPANESE GOVERMENT CITY & ENVIRONMENT PAVILION
b : Osaka-shi, Osaka
c : Dentsu, Tokyu Agency J.V.
d : Takanori Hamada
e : Takanori Hamada, Katsuhisa Ueda, Nobuo Nagasawa
f : Nomura Co.,Ltd.
g : Masami Daito
h : Afforestation, the important roles of plants and trees and the symbol of living things in a modern city were the main theme for this construction.
i : FRP plastic, acrylic plastic, wood, movable equipment
j : Nomura Co.,Ltd.

作品名
Project name

106 GINZA BLDG.	121	オークヒルズ香里	202	建築家が建てた建売住宅（朝霧住宅）	203
21世紀の安全をめざして	161	大阪市咲くやこの花館	247	高知市自由民権記念館	153
F.A.S.もとまち	77	大阪市立科学館	158	神戸市メリケンパークファッションモニュメント	222
GRAVITY	223	大阪南YMCA会館	112	神戸製鋼所ラグビー部クラブハウス	42
IMPインターナショナルマーケットプレイス	72	大妻女子大学	233	コーナーハウス ドゥ	66
ISS X 東光寺本堂新築工事	236	大野城市「まどか・ふれあい通り」擁壁レリーフ	230	国際花と緑の博覧会「会場全体サイン」	226
MYCAL本牧	68	大山道再生計画―弦巻・池尻	221	国際交流会館 芝蘭会館	50
NIKKI-SONY共同ビル	113	沖ビジネス本社ビル	117	国際花と緑の博覧会 政府苑屋外サイン	225
NKF川崎ビルディング	110	奥郷屋敷	140	国際花と緑の博覧会 政府苑「都市・環境」館	248
NTT POCKET PARK ねやがわ	162	屋久杉自然館	157	国立科学博物館サイエンス・シアター	159
PATIO DE ROBLE 大林組御影社宅	201	小倉ビル	122	駒沢ガーデンハウス	196
POLA札幌南1条ビル	70	オムロン中央研究所3号館	130	コンフォート岡山	203
SHIRAKAWA RESIDENCE	196	花王「清潔と生活」小博物館	151	サウスサイドコート	43
S邸	207,210	加古川駅前通商店街モール	174	佐世保四ヶ町商店街アーケードの方位盤	243
TNN放送センター社屋	125	笠岡市立カブトガニ博物館	159	サッポロビール千葉工場ゲストハウス	131
URBAN VEGA	203	カデンツア・ザ・タワー モデル・ルーム	200	サッポロビール北海道工場	132
アーチ岩田	228	神奈川県立金沢文庫	156	さぬき信用金庫本店	122
藍澤証券白樺研修センター	46	神奈川トヨタ本牧ショールーム カレリア	239	シエスタ伊豆高原	71
青山Mビル	103	カネディアン アカデミィ	148	自然との対話	223
秋田総合生活会館・美術館「アトリオン」	234	鎌倉河合ビル	197	芝浦スクエア	160,220
朝日新聞100周年記念モニュメント	221	軽井沢リゾートヴィラ4号館	45	芝浦スクエアモニュメント「帆風」	220
足利市環境装置「歴史から未来へ」	222	寒霞渓パブリックトイレ	244	十六銀行 研修所	146
アシックススポーツ工学研究所人材開発センター	136	神田ホリイビル	125	ショッパーズプラザ新浦安	60,61
愛宕原ゴルフ倶楽部 クラブハウス	53	北とぴあ科学館	154	白鳥センチュリープラザ	127
あつまろう緑の仲間たち	224	喫煙テーブル	231	新川崎三井ビルディング	119
アトラス長門	243	鬼怒川保養所「せせらぎ荘」	40	心斎橋筋商店街アーケード	173
我孫子市 鳥の博物館	155	気の博物館	155	新宿171ビル	118
アムナット スクエア	36	岐阜メモリアルセンター	126	新宿モノリス	123
アライアンス	121	京都リサーチパーク	105	新鳥羽水族館	84
アライカーボン 八代工場	133	銀座松屋リファイン120食品	86	新日鉱ビル	116
飯田ビル	197	グランエクレール市谷	200	新橋NHビル	120
池田山ゲストハウス	195	グランシティ大津ヶ丘	229	信用組合大阪興銀八尾支店	100
伊勢丹新宿店2階「ザ・メッセージ」	65	グランディ新横浜	74	スタンフォード日本センター	102
イヌイビル・カチドキ・サイン計画	225	グリーンホール相模大野サイン計画	227	スタンレー電気秦野製作所新2号館	135
岩鋳キャスティングワークス	64	クリスタルタワー	98	スパリゾートハワイアンズ・スプリングパーク	75
牛久自然観察の森	231	紅の家NO.1	206	スパリゾートハワイアンズ	75,228
ウッドピアいわき	156	ぐんまこどもの国児童会館	143	住友生命OBPプラザビル いずみホール	142
エクシブ白浜	82	京阪京橋駅ビル	62	西神そごう出店計画	76
エル・キャビン	205	京葉国際カントリークラブ	56	星和台ファミリーホール	128
大石田町クロスカルチャープラザ	139				

積水ハウス総合住宅研究所 ……………146
セコムHDセンター名張 ……………141
世田谷弦巻ハイム ……………197
第一ホテルアネックスサイン計画 ……………228
大栄教育システム名張研修所 ……………145
大同生命福井第2ビル ……………105
大同生命山形ビル ……………109
大本山總持寺三松閣 ……………236
大洋グループ軽井沢山荘 ……………210
太陽生命高知ビル ……………125
高岡市万葉歴史館 ……………157
武石村ともしび博物館 ……………158
嶽之下宮御社殿 ……………237
蓼科ブライトン倶楽部 ……………59
玉木病院 ……………148
多摩そごうシンボルゾーン「多摩の四季」……………85
チチヤス本社ビル ……………111
千葉市都市装置 ……………220
千葉中央銀座ふれあい商店街モニュメント計画 ……227
中外製薬株式会社宇都宮工場 ……………134
調布駅北口ビル(調布パルコ) ……………73
テクノウェイブ100新築工事 ……………19
デナマボール ……………184
東京写真専門学校四番町校舎 ……………150
東京生命本社ビル ……………108
東京セサミプレイス ……………80
東京ベイホテル東急 ……………38
東京よみうりカントリークラブ ……………54
東武スーパープール ……………78
トキメックビル ……………179
トヨタ・オートサロン アムラックス東京 ……………241
トヨタオート大阪サンダンス2000 ……………129
トヨタ自動車池袋ビル AMLUX ……………240
トヨタ博物館 ……………152
豊橋市自然史博物館 ……………151
名古屋商科大学コミュニティーセンター ……………181
西池袋公園 ……………183
西伊豆町総合ギャラリーパーク計画 …………22,180
日新製鋼株式会社呉製鉄所若葉研修センター ……147

西島プレス白井工場 ……………130
西天神社新築工事 ……………237
西宮観光30選説明サイン ……………226
日産アプリーテ千葉 ……………242
日産プリンス京都本社ショールーム ……………242
ニッセイ高知本町ビル ……………107
ニッセイ・ライフ・プラザ新宿 ……………106
日東紡ビルサイン計画 ……………227
ニドム リゾート ……………44
日本合成ゴム筑波研究所 ……………138
乃村工藝社大阪事業所 ……………161,162
乃村工藝社大阪事業所II役員室 ……………162
乃村工藝社大阪事業所III ……………162
白山オフィス ……………120
白馬コルチナリゾート・ヴィラ ……………49
バスストップの時計塔 ……………224
花園東公園 ……………183
花の木ゴルフクラブ ……………57
東広野ゴルフ倶楽部 ……………58
久屋大通公園 復興事業収束モニュメント ……………176
ビジネスホテル関屋 ……………34
日立基礎研究所 ……………137,194
日立基礎研究所 美山クラブ ……………194
ビッグワンカントリークラブ信楽コース ……………54
日時計 ……………222
日之影町立日之影中学校 ……………149
姫路城周辺地区総合サイン ……………226
広島市観光サイン ……………230
備後町山口玄ビル ……………124
福井県自然保護センター ……………143
福岡市博物館 ……………154,161
福岡タワー ……………172
福武書店本社ビル ……………101
株式会社フジタカ本社社屋 ……………99
芙蓉ミュージカルシアター ……………83
ブリックハウス・あしび ……………70
ベルクルーエ小諸 ……………209
法隆寺金堂壁画模写展示館 ……………238
北海道安達学園 ……………150

ホテルグリーンプラザ大阪アネックス ……………39
ホテルスプリング幕張 ……………35
ホテル日航福岡のプランター ……………39
ホテル日航 福岡 ……………184
ポラール一番館 ……………63
堀口邸 ……………208
毎日放送本社ビル ……………97
マキシーショッピングセンター・リニューアル ……………231
松下IMPビル ……………106
未完成ビル屋上テント増築 ……………244
三井倉庫箱崎ビル ……………18
三菱倉庫越前堀再開発 ……………20
見富邸 ……………210
南大津通商店街モール ピュアO2 ……………182
魅力ある道路づくり事業 区役所前通り ……………174
武藤工業本社ビル ……………124
室津海水浴場整備 ……………76
メイプルパークスクエア ……………199
めぐみ学園リフレッシュ工事 ……………150
もめんや藍 ……………243
山形生涯教育センター ……………156
ユーノス・サイン・システム ……………229
ユーラシア404 ……………47
横浜アリーナ ……………51
横浜伊勢佐木町ワシントンホテル ……………32
横浜そごう雲見迎賓館 ……………52
横浜博覧会テーマゾーン ……………245
横浜ビジネスパーク ……………114
代々木ゼミナール岡山校 ……………149
リバーテラス ……………204
レイント・ハウス 横浜 ……………178
レランドセンタービル ……………123
六甲アイランドCITYイーストコート4番街 ……………198
ロッテワールド・サイン計画 ……………232
ロワ・ヴェール荻窪北 ……………194
和歌山朝日ビル ……………115
和多田邸 ……………206
わんぱく天国 ……………183

ディレクター
Director

GK設計 ……………………22,103,174,180,230
YBP設計室 ……………………114
アール・アイ・エー ……………………60
相原慎一 ……………………156,157
秋山茂樹 ……………………229
朝倉則幸 ……………………22,180
朝永徹一 ……………………64
阿部和信 ……………………80
天川雅晴 ……………………76
飯笹正勝 ……………………179
猪狩晃一 ……………………231
池上俊郎 ……………………162,203,223
石川晃二 ……………………220
石阪春生 ……………………58
石田信行 ……………………236
石橋建築 ……………………162
石原堅次 ……………………157
石見修一 ……………………245
市川元則 ……………………76
稲垣栄三 ……………………236
井上富 ……………………157
猪倉啓行 ……………………198
岩尾エンジニヤリング ……………………230
植草孝 ……………………61
上田信也 ……………………97
上野卓二 ……………………78,137
上村彰雄 ……………………183
宇口進 ……………………138
後山悟 ……………………156
内田繁 ……………………86
エス・エフ・メーカーズ ……………………241
江幡正之 ……………………178
遠藤英雄 ……………………156,220
大熊俊隆 ……………………232
大野昭夫 ……………………157
大林組…42,43,63,66,102,105,107,114,124,125,128,196,201,202
大森文樹 ……………………151

岡田露愁 ……………………99
沖ディベロップメント ……………………117
奥田総一郎 ……………………222
折口紀子 ……………………183
学習研究社 環境企画設計室 ……………………231
蔭山利夫 ……………………150,161,162,200
梶川伸二 ……………………225,226
鹿島建設インテリアデザイン部 ……………………39
鹿島建設 ……………………39,78,80,134,137,178,179,184,194
角永博 ……………………157
金子太郎 ……………………65
加納隆 ……………………244
神尾勝宏 ……………………242
川瀬友造 ……………………151
神部千太郎 ……………………231
菊地正典 ……………………209,210
岸谷孝一 ……………………236
木谷靖孫 ……………………240
計量計画研究所 ……………………243
小池俊之 ……………………116
郷力憲治 ……………………72,161,162
國分孝雄 ……………………210
国立科学博物館 ……………………159
小嶋凌衛 ……………………125,149
小嶋敏男 ……………………197
近藤繁 ……………………82
坂倉建築研究所 ……………………32
佐々木桂 ……………………44
佐世保市 ……………………243
佐藤総合計画 ……………………161
佐藤孝幸 ……………………61
芝田良治 ……………………61,232
小角亨 ……………………146
杉本洋文 ……………………245
鈴木成文 ……………………236
清家清 ……………………157
船場 ……………………243
創和コーポレーション ……………………244

大成道路 ……………………227
高野裕之 ……………………156,157
高橋成忠 ……………………157
高橋信裕 ……………………157
高山順行 ……………………126,127
武井淳 ……………………243
武居清 ……………………159
竹中工務店…18,19,20,32,35,36,50,51,53,62,70,98,110,111,112,113,115,124,129,130,135,136,140,145,147,148,158,160,173,181,182,195,196,197,199,204,225,237,238,242,243,245
竹中工務店東京本店設計部 ……………………197
谷岡義高 ……………………71
茅野秀真 ……………………234
津田雅人 ……………………157
テイ・グラバー ……………………64,225,227,228
寺本敏則 ……………………106,153
電通 ……………………239
東京建物 ……………………122
東京大学 ……………………236
東畑 ……………………162
都市計画 ……………………77,174
友水孝一 ……………………74
中沢文三 ……………………154
中田文彦 ……………………205
中辻伸 ……………………85,156,221,222
中本嘉彦 ……………………39,45,49,150
中山隆 ……………………159
西澤健 ……………………22,103,174,180
日建設計…40,82,97,101,105,106,126,127,141,142,146,151,152,153,172,176,198,231,233,234,240,247
日本設計 ……………………38,75,116,119,123,138,227,228
乃村工藝社…44,61,65,72,85,86,150,156,157,159,161,162,200,220,221,222,227,229,232,239,241,243,248
野村不動産 ……………………114
橋本巌 ……………………101,105,247
長谷工コーポレーション ……………………59,60,71,205

デザイナー
Designer

花輪恒 ……………………………… 180
羽生昌弘 ………………………… 134,137
濱田隆法 …………………………… 248
林行正 …………………………… 143,155
原田豊 ……………………………… 157
伴田浩 ……………………………… 77
番晶勝 ……………………………… 122
日立建設 ………………………… 137,194
平田翰那 …………………………… 184
福井正和 …………………………… 207
福岡市 …………………………… 154,161
福家克彦 …………………………… 244
藤井明文 ………………………… 154,158
藤井栄一 …………………………… 106
堀田勝之 ………………………… 156,157
堀越裕二 …………………………… 120
増田通二 …………………………… 73
松川澄夫 …………………………… 86
松村数正 …………………………… 159
真鍋実 ……………………………… 161
萬川純一 …………………………… 76
南石周作 ………………………… 146,152,176
南和正 …………………………… 22,103
三橋一彦 …………………………… 139
三宅健司 …………………………… 40
宮沢巧 ……………………………… 226
宮武博彦 …………………………… 157
村岡賢二 …………………………… 229
望月昭 ……………………………… 157
本松浩司 …………………………… 155
森田耕二 …………………………… 99
森山晶夫 …………………………… 159
山口広喜 …………………………… 143
山口勝 ……………………………… 154
山本博 ……………………………… 172
横浜市都市計画局 ………………… 174
与謝野久 …………………………… 142
吉岡啓賀 …………………………… 210
吉川昭 ……………………………… 233

A.R.P.E社 ………………………… 38
GK設計 …………………… 22,103,174,180,230
J.フィッシャー …………………… 184
LOTTE蚕室建設本部 …………… 232
TLヤマギワ研究所 ……………… 184
YBP設計室 ……………………… 114
アーク・クルー …………………… 132
アーバンガウス研究所 ……… 162,203,223
合川彰夫 …………………………… 145
相原慎一 ………………………… 156,157
空充秋 ……………………………… 146
秋山哲也 …………………………… 108
浅井康行 …………………………… 36
葛原定次 ………………………… 126,127
東政美 ……………………………… 62
集山一廣 ………………………… 36,182
安部充 ……………………………… 54
有角昇純 ………………………… 18,32
安藤清 ……………………………… 36
伊井伸 ……………………………… 57
猪狩晃一 …………………………… 231
池上一夫 …………………………… 71
池上俊郎 ………………………… 162,203,223
池邊哲治 …………………………… 155
石井敏和 …………………………… 231
石井泰二 ………………………… 156,221,222
石井幹子 …………………………… 123
石倉義行 …………………………… 146
石坂浩一 …………………………… 61
石田信行 …………………………… 236
石野連司 …………………………… 35
磯村克郎 …………………………… 174
市川国夫 …………………………… 176
市川浩一 …………………………… 86

イデア造園設計 …………………… 184
井出昭治 …………………………… 114
伊藤崇洋 …………………………… 68
伊藤治磨 …………………………… 122
稲垣ランドスケープデザイン研究所 … 75
稲垣丈夫 …………………………… 75
猪子順 ……………………………… 146
稲地一晃 …………………………… 174
井上宜延 …………………………… 173
井上軍 …………………………… 111,147
井上隆夫 …………………………… 149
井上雅雄 …………………………… 124
井下公彦 …………………………… 143
今井彰 ……………………………… 65
今里清 …………………………… 122,194
稲葉利行 …………………………… 35
岩尾磁器工業 ……… 39,161,162,222,224,230,243
イワオ・タイルデザインチーム …… 161
イワオ・特販デザインチーム …… 230,243
イワオ・ベセラ・デザインチーム … 39,224
岩崎堅一 …………………………… 245
岩崎正 ……………………………… 65
岩下啓和 …………………………… 243
岩下龍吉 …………………………… 199
岩本勝也 ………………………… 154,158
上坂脩 ……………………………… 129
上田万寿 …………………………… 248
植田明 ……………………………… 243
上野教人 ………………………… 130,135
上野卓二 ………………………… 78,137
上野美子 …………………………… 162
牛山恭男 …………………………… 114
臼井真 ……………………………… 32
薄田学 ……………………………… 50

歌一洋 ……………………………………208,231
歌田次男 ………………………………………59
宇田靖夫 ………………………………………221
内海利彦 ………………………………………100
梅原二六 ………………………………………59
浦山哲 …………………………………………54
榎本和夫 ………………………………………51
遠藤英雄 ……………………………………156,220
及川義邦 ………………………………………20
大萱喜知郎 ……………………………………157
大下稔夫 ………………………………………243
大達智子 ………………………………………86
大塚慶子 ………………………………………201
大友昭治 ………………………………………220
大野一男 ………………………………………153
大橋成実 ………………………………………232
大橋眞由美 …………………………………63,128
大湊浩幸 ………………………………………226
大林組… 42,43,63,66,102,105,107,114,124,125,128,
　　　196,201,202
大林組設計部 …………………………………201
大林組本店設計部 ……………………………196
小笠原祥伸 ………………………………46,197,203
岡田露愁 ………………………………………99
岡田哲哉 ………………………………………72
小川清一 ………………………………………148
荻原剛 …………………………………………245
奥田総一郎 ……………………………………222
小澤浩利 …………………………………143,151
小沢真由美 ……………………………………183
小城由加子 ……………………………………232
尾関勝之 ………………………………………114
織田誠爾 ………………………………………232
小野里宏 ……………………………………183,229
小野為一 ………………………………………106

小野寺康夫 ………………80,134,137,179,194
折笠智和 ………………………………………220
海宝幸一 ………………………………………240
貝山秀明 ……………………………………183,229
鍵山一生 ………………………………………74
影山温子 ………………………………………183
梶川伸二 ……………………………………225,226
鹿島建設………39,78,80,134,137,178,179,184,194
柏木浩一 ……………………………………115,136
梶原一幸 ………………………………………43
片岡麻子 ………………………………………155
片山賢 …………………………………………231
加藤園子 ………………………………………244
加藤宏生 ………………………………………36
門川清行 ………………………………………145
角島健二 ………………………………………130
金沢義人 ………………………………………196
金山正志 ………………………………………44
金子清二 ……………………………85,156,220,221
加納建築設計事務所 …………………………244
亀谷美幸 ………………………………………180
刈屋珠明 ………………………………………159
狩野忠正 …………………………129,140,173
川勝英史 ……………………………………239,241
河崎康了 ………………………………………182
川崎昭夫 ………………………………………83
川瀬俊二 ………………………………………114
川畑英幹 ………………………………………155
川村光明 ………………………………………145
川村清志 ……………………………………61,232
河村光男 ………………………………………68
環境開発研究所 ………………………………158
神田孜 ……………………………………130,135
菊田靖久 ………………………………………106

菊地章 ……………………………………209,210
菊池東紀男 ……………………………………138
木島幸一 ………………………………………160
木田輝夫 ………………………………………44
木田和秀 ………………………………………61
北川寿雄 ………………………………………75
北島正行 ………………………………………198
木戸康裕 ………………………………………154
絹川正 …………………………………………160
究建築事務所 …………………………………236
桐岡栄 …………………………………………243
切道敏郎 ………………………………………61
桐山昇一 ………………………………………155
近代造形 ………………………………………229
金納弘一 ………………………………………73
空間計画研究所 ………………………………59
櫛谷諒 …………………………………………231
倉田明夫 ………………………………………200
倉谷寛 …………………………………………242
倉光瑞枝 ………………………………………61
栗原幸二 ………………………………………201
栗原礼子 ………………………………………159
黒田幸夫 ………………………………………184
黒田信行 ………………………………………194
黒野雅好 ………………………………………138
桑名正一 ………………………………………121
毛塚洋 …………………………………………114
建築委員会 …………………………………161,162
甲田敦子 ………………………………………154
河南誠 …………………………………………125
河野晴彦 ……………………………………117,118
國分孝雄 ………………………………………210
五戸准 …………………………………………200
奥石勲 …………………………………………84

小嶋凌衛	125,149	柴田嘉夫	82	高野菊男
小嶋敏夫	195	嶋俊己	237	高橋昭伸
児島正剛	85	清水弘之	145	高山稔
小菅敏彦	84	清水一夫	68	武石正宣
児玉慎憲	137,178,179	東海林孝司	131	竹井俊介
児玉もえみ	137,179	碇屋雅之	131	竹尾健一
後藤浩美	148	白木孝志	53	武田博宣

小嶋凌衛 ‥‥‥‥‥‥‥‥‥‥‥125,149
小嶋敏夫 ‥‥‥‥‥‥‥‥‥‥‥‥195
児島正剛 ‥‥‥‥‥‥‥‥‥‥‥‥85
小菅敏彦 ‥‥‥‥‥‥‥‥‥‥‥‥84
児玉慎憲 ‥‥‥‥‥‥137,178,179
児玉もえみ ‥‥‥‥‥‥‥137,179
後藤浩美 ‥‥‥‥‥‥‥‥‥‥‥148
小林泰樹 ‥‥‥‥‥‥‥‥‥111,147
西条由治 ‥‥‥‥‥‥‥‥‥121,200
酒井利行 ‥‥‥‥‥‥‥‥‥115,136
坂口彰 ‥‥‥‥‥‥‥‥‥‥‥‥130
坂田克哉 ‥‥‥‥‥‥‥‥‥‥‥245
坂根恭司 ‥‥‥‥‥‥‥‥‥‥‥‥53
桜井篤信 ‥‥‥‥‥‥‥‥‥‥‥‥47
佐々木賢範 ‥‥‥‥‥‥‥‥22,180
佐藤工業 ‥‥‥‥‥‥‥‥‥‥‥‥19
佐藤正二 ‥‥‥‥‥‥‥‥‥‥‥148
佐藤積義 ‥‥‥‥‥‥‥‥‥‥‥‥36
佐藤総合計画 ‥‥‥‥‥‥‥‥‥161
佐藤誠 ‥‥‥‥‥‥‥‥‥‥‥‥197
佐藤正徳 ‥‥‥‥‥‥‥‥‥‥‥‥44
佐藤真由美 ‥‥‥‥‥‥‥‥‥‥232
真田元晴 ‥‥‥‥‥‥‥‥‥‥‥150
佐野隆夫 ‥‥‥‥‥‥‥‥‥‥‥108
佐溝裕紀 ‥‥‥‥‥‥‥‥‥‥‥122
澤田勝 ‥‥‥‥‥‥‥‥‥‥38,195
椎名清作 ‥‥‥‥‥‥‥‥‥‥‥‥61
塩井保則 ‥‥‥‥‥‥‥‥‥‥‥247
潮先克 ‥‥‥‥‥‥‥‥‥‥‥‥‥62
塩津一興 ‥‥‥‥‥‥‥‥‥‥‥132
塩原敏男 ‥‥‥‥‥‥‥‥‥‥‥‥32
繁野舜 ‥‥‥‥‥‥‥‥‥‥‥‥182
執行昭彦 ‥‥‥‥‥‥‥‥‥156,157
地主道夫 ‥‥‥‥‥‥‥‥‥113,195

柴田嘉夫 ‥‥‥‥‥‥‥‥‥‥‥‥82
嶋俊己 ‥‥‥‥‥‥‥‥‥‥‥‥237
清水弘之 ‥‥‥‥‥‥‥‥‥‥‥145
清水一夫 ‥‥‥‥‥‥‥‥‥‥‥‥68
東海林孝司 ‥‥‥‥‥‥‥‥‥‥131
碇屋雅之 ‥‥‥‥‥‥‥‥‥‥‥131
白木孝志 ‥‥‥‥‥‥‥‥‥‥‥‥53
白浜力 ‥‥‥‥‥‥‥‥22,103,180
鈴木恵宜 ‥‥‥‥‥‥‥‥‥‥‥‥20
鈴木俊幸 ‥‥‥‥‥‥‥‥‥85,156
鈴木裕美 ‥‥‥‥‥‥‥‥‥‥‥122
スタジオ・ビー ‥‥‥‥‥‥223,224
須藤純治 ‥‥‥‥‥‥‥‥‥‥‥‥65
砂川裕幸 ‥‥‥‥‥‥‥‥‥‥‥140
澄川喜一 ‥‥‥‥‥‥‥‥‥‥‥123
住友林業緑化 ‥‥‥‥‥‥‥‥‥141
関原聡 ‥‥‥‥‥‥‥‥‥‥‥‥‥40
セサミジャパン ‥‥‥‥‥‥‥‥‥80
千本木隆芳 ‥‥‥‥‥‥‥‥‥‥236
千名良樹 ‥‥‥‥‥‥‥‥‥‥‥143
荘司訓由 ‥‥‥‥‥‥‥‥‥‥‥‥61
ダイエーリアルエステート ‥‥‥‥‥60
ダイエーリアルエステート浦安内装監理室 ‥‥‥60
大成建設設計本部‥‥ 34,46,47,52,54,56,68,70,73,83,
　　　108,109,117,118,120,121,122,
　　　123,125,131,132,145,149,194,
　　　197,200,203
大成建設大阪支店設計部 ‥‥‥‥54,58,100
大成建設札幌支店設計部 ‥‥‥‥‥‥47
大成建設四国支店設計部 ‥‥‥‥‥122
大成建設名古屋支店設計部 ‥‥‥‥57,84
高柴牧子 ‥‥‥‥‥‥‥‥‥‥‥106
高須芳史 ‥‥‥‥‥‥‥‥‥‥‥197
高瀬昭男 ‥‥‥‥‥‥‥‥‥‥‥183
高野勝也 ‥‥‥‥‥‥‥‥‥‥‥240

高野菊男 ‥‥‥‥‥‥‥‥‥‥‥229
高橋昭伸 ‥‥‥‥‥‥‥‥‥‥‥156
高山稔 ‥‥‥‥‥‥‥‥‥‥‥‥229
武石正宣 ‥‥‥‥‥‥‥‥‥‥‥‥61
竹井俊介 ‥‥‥‥‥‥‥‥‥‥‥‥47
竹尾健一 ‥‥‥‥‥‥‥‥‥‥61,232
武田博宣 ‥‥‥‥‥‥‥‥‥‥‥156
竹中工務店‥‥ 18,19,20,32,35,36,50,51,53,62,70,98,
　　　110,111,112,113,115,124,129,130,135,
　　　136,140,145,147,148,158,160,173,181,
　　　182,195,196,197,199,202,204,225,237,
　　　238,242,243,245
竹中工務店東京本店設計部 ‥‥‥‥‥197
竹林善孝 ‥‥‥‥‥‥‥‥‥‥‥226
竹本明美 ‥‥‥‥‥‥‥‥‥‥‥200
田島綏夫 ‥‥‥‥‥‥‥‥‥‥‥245
田代廣信 ‥‥‥‥‥‥‥‥‥‥‥110
建入通 ‥‥‥‥‥‥‥‥‥‥‥‥‥61
田中勝男 ‥‥‥‥‥‥‥‥‥‥‥152
田中孝次 ‥‥‥‥‥‥‥‥‥‥‥245
田中弘成 ‥‥‥‥‥‥‥‥107,201,202
田中政明 ‥‥‥‥‥‥‥‥‥‥‥243
田辺詔二郎 ‥‥‥‥‥‥‥‥‥‥120
田辺真知子 ‥‥‥‥‥‥‥‥‥‥198
谷口望 ‥‥‥‥‥‥‥‥‥‥‥‥141
谷田訓子 ‥‥‥‥‥‥‥‥‥‥‥239
田平徹 ‥‥‥‥‥‥‥‥‥‥‥‥‥47
田村賢治 ‥‥‥‥‥‥‥‥‥‥22,180
俵雅人 ‥‥‥‥‥‥‥‥‥‥208,231
千葉修良 ‥‥‥‥‥‥‥‥‥‥‥‥64
茶谷明男 ‥‥‥‥‥‥‥‥‥‥‥‥51
中央設計 ‥‥‥‥‥‥‥‥‥‥‥243
千代田化工建設 ‥‥‥‥‥‥‥‥‥19
塚田哲也 ‥‥‥‥‥‥‥‥‥‥‥109
塚本武司 ‥‥‥‥‥‥‥‥‥‥‥‥44

津路善巳 …………………………………198
土屋勝彦 …………………………………232
テイ・グラバー …………………64,225,227,228
手塚清次郎 ………………………………132
寺岡俊彦 …………………………………101
寺島振介 ……………………………78,134
東京セサミプレイス ……………………80
峠孝治 ……………………………………173
徳本幸雄 …………………………………32
都市計画 ……………………………77,174
戸塚元雄 …………………………………244
富家大器 …………………………………225
冨田徹 ……………………………………181
富田博 ……………………………………58
富松太基 …………………………………116
豊田幸夫 …………………………………178
トヨタ自動車 ………………………152,240
豊村博 ……………………………………113
中井進 ……………………………………97
永井英敏 …………………………………54
永井幹子 ……………………………71,205
中川千早 …………………………………146
長沢信夫 …………………………………248
永島亮太郎 ………………………………245
長嶋達也 …………………………………149
中島猛夫 …………………………………157
中田文彦 …………………………………205
中辻伸 ………………………85,156,221,222
中西博隆 …………………………………142
中村勲 ……………………………………242
中村和久 ……………………………122,194
中村一 ……………………………………146
中村秀男 …………………………………74
中村雅人 ……………………………231,233

中村裕輔 …………………………………72
長本朝子 …………………………………244
梨田俊次 ………………………39,49,150
西崎徳房 …………………………………36
西澤健 …………………………22,103,174,180
西田正則 ……………………………40,172
西田幸男 …………………………………70
西牟田章夫 ………………………………52
西村敏信 …………………………………156
西本博 ……………………………………181
西山千明 ……………………………61,232
西山慎一郎 ………………………………76
日建設計…40,82,97,101,105,106,126,127,141,142,
146,151,152,153,172,176,198,231,233,
234,240,247
日建ハウジング …………………………198
二宮孝 ……………………………………51
日本カラーテクノロジー研究所 …………206
日本設計 …………38,75,116,119,123,138,227,228
日本造園設計 ……………………………138
能丸弘毅 …………………………………202
野沢克昌 …………………………………159
野田隆史 …………………………………140
野中建築事務所 …………………………133
乃村一級建築士事務所 ……………161,162
乃村工芸社 ………………………………60
乃村工藝社…44,61,65,72,85,86,150,156,157,159,
161,162,200,220,221,222,227,229,232,
239,241,243,248
野村不動産 ………………………………114
野村充 ………………………………50,130
荻野久男 …………………………………86
橋本博 ……………………………………245
橋本緑郎 ………………………121,123,125,197
長谷工コーポレーション ………59,60,71,205

長谷工コーポレーションエンジニアリング事業部 ……59,
60,71,205
長谷工 ……………………………59,60,71,205
服部紀和 …………………………………113
花田佳明 …………………………………142
羽生昌弘 ……………………………134,137
濱口倫壽 …………………………………84
濱田隆法 …………………………………248
早川典臣 ……………………………225,226
林浩二 ……………………………………43
原尚正 ……………………………………229
原田昭久 ……………………………223,224
原田文夫 …………………………………85
春田義行 …………………………………56
伴泉 ………………………………………156
久林正晴 ……………………………18,242
日立建設 ……………………………137,194
人見修 ……………………………………68
日比野万喜男 ……………………………151
平野正治 …………………………………231
平部喜弘 …………………………………226
平山郁夫 …………………………………157
広島正明 …………………………………242
広田和正 ……………………………102,125
ヴィンチェンツォ・イアヴィコリ …………106,234
福井正和 …………………………………207
福島徹 ……………………………………159
福島勇人 …………………………………86
福田直人 …………………………………158
福田吉宏 …………………………………241
福地誠 ……………………………………77
福山秀親 …………………………………72
藤井昭造 …………………………………146
藤井明文 ……………………………154,158

フジタ建設	60	三木重典	148	矢島峰雄	105
藤田和孝	172	三木正	78	柳学	116
藤田雅俊	103	右高良樹	36	柳澤ゆき	143,151,154
藤縄正俊	42,63,66,105,124,125,128	三谷幸司	43,102,196	八幡健志	153
藤野和男	139	三井富雄	182,237	薮内秀美	223
藤本聖徳	143	皆神憲雄	86	山口美弘	155
藤森宣光	143,161	南裕二郎	196	山口一平	109,117,118,121,122,123,125,194,197
船津茂	72	美濃吉昭	54,58,100	山下祥三	199
船橋昭三	42,63,66,105,124,125	三橋一彦	139	山田宗彦	107
古田博司	181	三村拓也	72	山田達行	108
プロダクト翔	75	梅沢誠一郎	159	大和測量	222
ポール・マランツ	184	宮治長一	202	山元弘和	98
保川彰宏	72	宮田勝弘	110	山本範人	82
保科朝男	56	宮武仁	238	山森博之	161
星野一	105	宮田正人	228	由比清隆	150
細田理枝子	86	宮野哲也	157	有居徹彦	150,200
堀口壮平	196	宮野裕	243	由井真波	225,226
堀越裕二	120	宮原義治	132	指田茂	155
本城邦彦	112	三好恵介	157	横塚潔	61,232
本多友常	98	三輪祐児	156	吉岡一郎	238
本間利雄	107	向隆宏	85,156,220,222	吉岡啓賀	210
前川清	148	村上元康	62	吉川泰次	39,45
前田俊雄	112	元井正太郎	150	吉澤久枝	151
前田紀貞	34	森井浩一	162	吉田進	73
前田正英	200	森岡誠	226	吉田孝	202
前田康貴	131	森川和法	156	古田智二	238
正木素行	194	森川昇	51	吉田昇	135
枡野俊明	138	森繁文雄	68	吉田寛史	204
松岡俊之	110	森田昌嗣	174	吉村祐子	150
永松照基	76	モリチュウ	229	吉村佳哲	76
松永博巳	103,180	森昌子	83	吉本浩	45,49
松浦恒久	156	森山純治	154	ロータリーコーポレーション	243
マリア・ルイザ・ロッシ	106,234	矢口進	20,124	若林亮	176
丸山茂	146	横山芳祝	202	渡辺康人	206

撮影者
Photographer

KEN青山写真事務所 ……………………36
SS大阪 ………………43,54,105,107,202
SSグループ ……………………………150
SS現像所 …………………………………36
SS東京 ………32,108,121,134,137,179,194
SS東北 …………………………………234
SS名古屋 …………126,127,146,151,176
STUDIOさわだ …………………………38
T.ナカサ・アンド・パートナーズ ……103,174,180
Work Shop ……………………………229
アーキフォトKATO ……………………237
アーバンガウス研究所 ……162,203,223
合田建築写真研究所 ……………………122
青山写真館 ………………………………243
安達治 ……………………………………47
アトリエ フクモト ……………………242
安部順 ………………………………64,228
イエローフラッグ ………………………100
池上俊郎 ……………………162,203,223
石井敏和 …………………………………231
石田信行 …………………………………236
伊藤直明 …………………………………133
稲住泰広 ……………………………82,106
井上和幸 ……………………………184,222
井上光伸 …………………………………73
岩崎和幸 …………………………………38
上谷重男 …………………76,203,206,237
上諸尚美 ……………………………105,124
歌一洋 ………………………………208,231
エイトプロダクション …………………44
大島勝寛 ……………………62,158,199
大嶋勝寛 …………………………………129
岡田泰治 …………………………………200
尾崎保雄 …………………………………237
小野里宏 ……………………………183,229
小原勉 …………143,154,158,159,161
織田努写真事務所 ………………………63
貝山秀明 ……………………………183,229
鹿島建設 ………39,78,80,134,137,178,179,184,194
加藤敏明 …………………………………237
鏑木宏 ……………………………………197
カブラギ写真事務所 ……………………74,197
カブラギスタジオ ………………………38
カブラギ写真事務所 ……………………74,197

上村彰雄 …………………………………183
上諸尚美 ……………………………105,124
柄松稔 ……………………………………106
狩野正和 …………………………………123
カルダン・プロ …………………………243
川澄建築写真事務所 …116,119,123,137,138,178,179,184
川本斉 ……………………………………232
関西カラー写真 …………………………143
菅野哲也 ……………………………75,109
北沢治夫 …………………………………121
稲地一晃 …………………………………174
絹巻豊 ……………………………………200
木下保 ……………………………………244
共同プロ …………………………………247
クリエイティブ・スタジオ・カワカミ …179
車田写真事務所 ……………………182,238
車田保 ………………………………182,238
黒田青巌 ……………………………142,153
輿水進 ………………………225,227,228
国際写真 ………………111,147,148,203
小嶋凌衛 ……………………………125,149
後藤秀雄 …………………………………84
斉藤さだむ …………………22,180,225,226
斎部功 ……………………………………160
佐竹勝也 …………………………………49
佐田山靖雄 …………………………220,221
佐藤喜之 …………………………………86
澤田勝良 …………………………………38
ジー・バイ・ケー ………………………234
篠澤建築写真事務所 ……………………34
島修一 ……………………………………102
清水昭 ……………………………………178
水写真事務所 ……………………………178
ジャプロ …………………………………61
住宅・都市整備公団 ……………………229
首藤将夫 ……………………………151,154,155
庄野啓 ………………………………98,112
常磐興産 …………………………………75
照屋 ………………………………………206
白鳥美雄 …………………………………114
伸和実業 …………………………………58
菅野哲也 ……………………………75,109
菅野哲也写真事務所 ……………………75

スタジオ・ベン ……………………158,159
スタディオ クリオ ……………………105
設計室アーチザン ………………………228
センター・フォト ………………………84
センターフォート ………………………152
センター・ホト …………………………57
そあスタジオ ……………………………244
大宣 ………………………………………149
大成建設広報部 …………………………83
大東正巳 …85,106,156,157,159,161,162,220,221,222,239,241,248
高井深 ……………………………………54
高崎清治 …………………………………132
高橋裕嗣 …………………………………198
竹中工務店 …18,19,20,32,35,36,50,51,53,62,70,98,110,111,112,113,115,124,129,130,135,136,140,145,147,148,158,160,173,181,182,195,196,197,199,202,204,225,237,238,242,243,245
竹村写真事務所 …………………………125
立見治嗣 …………………………………206
田中昌彦 …………………………………182
タミ・アート ……………………………130
東京D.P.E ………………………………71
東京グラフィック …………………20,124,197
東出清彦 ………43,62,66,97,140,146
長尾純之助 ………………………………203
中田文彦 …………………………………205
ナトリ光房 …………………42,128,130,196
新津写真 …………………………………70
野口写真事務所 …………………………120
西日本写房 …………………………101,172
野口毅 ………………117,121,123,125
服部寿徳 ……………………………122,194
ハットリスタジオ ……………………122,131,194
花城辰男 …………………………………107
原田昭久 ……………………………223,224
東出清彦 ………43,62,66,97,140,146
ヒロフォトビルディング ………………19
フォトセンター …………………………46
フォトハウス田中 ………………………182
深見耕一 …………………………………229
福本正明 …………………………………242
藤江猛 ……………………………………46
富士ゼロックス …………………………145

藤田幹夫 ……………………………………242
藤戸充 ……………………………………………64
古川泰造 …………………………………145,173
古田雅文 …………………………………53,204
細間邦明 …………………………………………207
堀内広治 ………………………………………245
牧明夫 …………………………………………196
牧野裕 …………………………………………244
マキ・フォトグラフィック ……………………196
松田博史 …………………………………209,210
マツダ・プロ・カラー …………………113,195
マツダ・プロカラー ……………………………240
松村芳浩 ……………………………………208
三沢ユタカ ……………………………………100
溝内了介 ……………………………………78,80
三橋一彦 ………………………………………139
ミヤガワ …………………………………35,242
三輪晃久写真研究所 ……………………197,236
三輪晃士 …………………………56,70,118,197
三輪写真 …………………………………47,68,210
三輪写真事務所 ………………………………210
三輪哲士 …………………………………………52
村井修 …………………………36,50,136,148,181
村岡正巳 ………………………………………203
村瀬武男 …………………………………72,150
名執一雄 …………………………………130,201
森下友二 …………………………………………99
門馬金昭 …………………………………40,197
ヤジマカメラ ……………………………………59
安川千秋 …………………………………22,180
山田商会 ……………………………………132
やまだ商会 …………………………………150
山本圭一 ……………………………………226
吉岡啓賀 ……………………………………210
吉川健児 …………………………………………39
吉倉光輝 …………………………………………77
吉田幸世 …………………………………………65
吉村行雄 …………………………70,113,135,145
米倉写真事務所 ………………………………120
和木通 ………………………………………240
渡辺敏正 ……………………………………230
渡辺洋美 …………………………………………51
ミヤガワ …………………………………35,242

GK大阪 ……………………………………226
GK京都 ……………………………………225,226
GK設計 ………………………22,103,174,180
ISS建築設計事務所 ……………………………236
YBP設計室（野村不動産＋大林組）………114
アーク造園設計事務所 …………………183,229
アルス …………………………………………76
池上俊郎＋アーバンガウス研究所 ………203,223
池上俊郎＋アーバンガウス研究所 ……………162
岩尾磁器工業 ……………39,161,162,222,224,230,243
上谷重男＋U建築企画室 ………76,203,206,237
歌一洋建築研究所 ………………………208,231
エー・アンド・エー ………………39,45,49,150
大林組…42,43,63,66,102,105,107,124,125,128,196,
　　　201,202
大林組本店建築設計第1部 ……………………43
大林組本店建築設計第5部 ……………………202
オリエントI級建築士事務所 ……………………207
鹿島建設………39,78,80,134,137,178,179,184,194
加納建築設計事務所 …………………………244
究建築事務所 …………………………………236
計画工房DNA …………………………………174
国分設計 ………………………………………210
小嶋凌衛建築設計事務所 …………………125,149
サカエ ……………………………184,222,223,224
サン吉岡建築設計事務所 ……………………210
シルフ …………………………………………99
新日本製鉄君津製鉄所 ………………………206
慧匠社建築研究所 ……………………………244

スペースデザイン ハッピー …………………231
設計室アーチザン ……………………………228
船場 ………………………………………242
像建築設計事務所 ………………………………74
大昌工芸 ……………………………………230
大成建設設計本部…34,46,47,52,54,56,57,58,68,70,
　　　73,83,84,100,108,109,117,118,
　　　120,121,122,123,125,131,132,
　　　145,149,194,197,200,203
竹中工務店…18,19,20,32,35,36,50,51,53,62,70,98,
　　　110,111,112,113,115,124,129,130,135,
　　　136,140,145,147,148,158,160,173,181,
　　　182,195,196,197,199,204,225,237,238,
　　　242,243,245
丹青社 ……………………120,143,154,158,159,161
丹青社オフィスデザイン研究所 ………………120
ディエム・デザイン ………………………209,210
テイ・グラバー …………………64,225,227,228
都市計画 ………………………………………77
日建設計…40,82,97,101,105,106,126,127,141,142,
　　　146,151,152,153,172,176,198,231,233,
　　　234,240,247
日展 ……………………………143,151,154,155
日本設計…38,75,116,119,123,138,227,228
野中建築事務所 ………………………………133
乃村工藝社…44,61,65,72,85,86,150,156,157,159,
　　　161,162,200,220,221,222,227,229,232,
　　　239,241,243,248
長谷エコーポレーション …………59,60,71,205
三橋建築設計事務所 …………………………139

株式会社アーク造園設計事務所
URC Landscape Design Office
〒160 東京都新宿区四谷2-8クローバービル2F
2F, Clover-Bldg., 2-8, Yotsuya, Shinjuku-ku, Tokyo, 160
Tel.03-3359-5049

設計室アーチザン
Artisan Architectural Design Office
〒440 愛知県豊橋市旭本町35
35, Asahihon-machi, Toyohashi-shi, Aichi, 440
Tel.0532-55-1002

ISS 建築設計事務所
ISS Architects & Engineers Associates
〒662 兵庫県西宮市柳本町8-22-701
8-22-701, Yanagihoncho, Nishinomiya, Hyogo, 662
Tel.0798-74-0633

株式会社アルス
Arusu Co.,Ltd.
〒550 大阪府大阪市西区立売堀1-9-36
1-9-36, Itachibori, Nishi-ku, Osaka-shi, Osaka, 550
Tel.06-531-9222

池上俊郎＋アーバンガウス研究所
Toshiroh Ikegami+Urban Gauss
〒550 大阪府大阪市西区京町堀1-8-31安田ビル310
310, Yasuda-Bldg., 1-8-31, Kyomachibori, Nishi-ku, Osaka-shi, Osaka, 550
Tel.06-448-8580

岩尾磁器工業株式会社
Iwao Jiki Kogyo Co.,Ltd.
〒884 佐賀県西松浦郡有田町1288番地
1288, Arita-machi, Nishi-Matsuura-gun, Saga, 884
Tel.0955-43-2111

上谷重男＋U 建築企画室
Shigeo Uetani+U Architectural Office
〒651 兵庫県神戸市中央区浜辺通4-1-23 三宮ベンチャービル312
312, Sannomiya Venture Bldg., 4-1-23, Hamabedori Chuo-ku, Kobe-shi, Hyogo, 651
Tel.078-222-6036

歌一洋建築研究所
Ichiyo Uta Architect & Associates
〒542 大阪府大阪市中央区南船場4-10-25第3飯沼ビル4F
4F, Daisan Iinuma Bldg., 4-10-25, Minami-Senba, Chuo-ku, Osaka-shi, Osaka, 542
Tel.06-252-4772

株式会社エー・アンド・エー
A & A Co.,Ltd.
〒543 大阪府大阪市天王寺区国分町17番21号
17-21, Kokubu-machi, Tennoji-ku, Osaka-shi, Osaka, 543
Tel.06-779-0151

株式会社大林組本店建築設計第1部
Obayashi Corporation, Design Departments, No.1
〒540 大阪府大阪市中央区北浜東4-33
4-33, Kitahamahigashi, Chuo-ku, Osaka-shi, Osaka, 540
Tel.06-946-4718

株式会社オリエント1級建築士事務所
Orient Architects & Planners
〒064 北海道札幌市中央区宮の森2条11-6-33
11-6-33, 2jo, Miyanomori, Chuo-ku, Sapporo-shi, Hokkaido, 064
Tel.011-642-6420

鹿島建設株式会社
Kajima Corporation
〒107 東京都港区赤坂6-5-30
6-5-30, Akasaka, Minato-ku, Tokyo, 107
Tel.03-5561-2111

株式会社加納建築設計事務所
Kanoh Architectural Office
〒460 愛知県名古屋市中区千代田1-4-15
1-4-15, Chiyoda, Naka-ku, Nagoya-shi, Aichi, 460
Tel.052-263-9780

究建築事務所
Kyu Architectural Design Office
〒108 東京都港区高輪1-5-33-606
1-5-33-606, Takanawa, Minato-ku, Tokyo, 108
Tel.03-3449-0606

計画工房 DNA
Keikaku Kobo DNA
〒650 兵庫県神戸市中央区栄町通1-2-1住友信託6F
6F, Sumitomo-Sintaku, 1-2-1, Sakae-cho-dori, Chuo-ku.Kobe-shi, Hyogo, 650
Tel.078-391-0294

慧匠社建築研究所
Keishosha Archi-Tect
〒760 香川県高松市通町1-2
1-2, Torimachi, Takamatsu-shi Kagawa, 760
Tel.0878-22-7097

株式会社国分設計 代表取締役國分孝雄
Kokubu Sekkei Co.,Ltd. Architect TaKao Kokubu
〒460 愛知県名古屋市中区東桜2-22-18日興ビル7階
7F Nikko Bldg., 2-22-18, Higashisakura, Naka-ku, Nagoya-shi, Aichi, 460
Tel.052-931-7771

小嶋凌衛建築設計事務所，小嶋凌衛
Ryoe Kojima and Associates
〒882 宮崎県延岡市出北1-34-18
1-34-18, Idekita, Nobeoka-shi, Miyazaki, 882
Tel.0982-34-1128

株式会社サカエ
Sakae Co.,Ltd.
〒181 東京都三鷹市新川4-4-9
4-4-9, Shinkawa Mitaka-shi, Tokyo, 181
Tel.0422-47-5981

サン吉岡建築設計事務所
San Yoshioka Architectural Office
〒194 東京都町田市本町田1751-7
1751-7, Honmachida, Machida-shi, Tokyo, 194
Tel.0427-26-7064

株式会社シルフ
SYLPH
〒603 京都府京都市北区紫野宮東町10-3-203
10-3-203, Murasakinomiya-Higashi-cho, Kita-ku, Kyoto-shi, Kyoto, 603
Tel.075-432-2722

新日本製鉄株式会社君津製鉄所
Nippon Steel Corporation Kimitsu Works
〒299-11 千葉県君津市君津1番地
1, Kimitsu, Kimitsu-shi, Chiba, 299-11
Tel.0439-55-9076

株式会社 GK 設計
GK Sekkei Inc.
〒171 東京都豊島区南池袋1-11-22 山種ビル
Yamatane Building, 1-11-22, Minami-Ikebukuro, Toshima-ku, Tokyo, 171
Tel.03-3989-9511

株式会社 GK 大阪
GK Osaka Co.,Ltd.
〒541 大阪府大阪市中央区本町4-4-24 住友生命第2ビル
Sumitomo Seimei Dai2 Building, 4-4-24, Honmachi, Chuo-ku, Osaka-shi, Osaka, 541
Tel.06-252-2066

株式会社 GK 京都
GK Kyoto Co.,Ltd.
〒602 京都府京都市上京区相国寺東門前町657
657, Higashi-Monzenmachi, Sokoku-ji, Kamigyo-ku, Kyoto-shi, Kyoto, 602
Tel.075-211-2277

●
スペースデザイン ハッピー
Space Degin Happy
〒124 東京都葛飾区東新小岩3-11-6 第一大雄マンション205号
205, Daiichi-Daiyu Mansion, 3-11-6, Higashi-Shinkoiwa Katsushika-ku, Tokyo, 124
Tel.03-5698-1323

●
株式会社 船場
Semba Corporation Design Section
〒530 大阪府大阪市北区1-5
1-5, Kita-ku, Osaka-shi, Osaka, 530
Tel.06-313-1008

●
株式会社像建築設計事務所
Zou, Architects & Associates, Inc.
〒231 神奈川県横浜市中区松影町2-7-20福地ビル
Fukuchi Bldg., 2-7-20, Matsukage-Cho, Naka-ku, Yokohama-shi, Kanagawa, 231
Tel.045-641-8008

●
大昌工芸株式会社
Taisho Kogei Co.,Ltd.
〒733 広島県広島市西区小河内町2-15-2
2-15-2, Ogawachi-machi Nishi-ku, Hiroshima-shi, Hiroshima 733
Tel.082-295-6922

●
大成建設株式会社設計本部
Taisei Corporation Design and Proposal Division
〒163 東京都新宿区西新宿1-25-1新宿センタービル
Shinjuku Center Bldg.1-25-1, Nishi-Shinjuku, Shinjuku-ku, Tokyo, 163
Tel.03-3348-1111

●
株式会社竹中工務店
Takenaka Corporation
〒541 大阪府大阪市中央区本町4-1-13
4-1-13, Honcho, Chuo-ku, Osaka-shi, Osaka, 541
Tel.06-252-1201

●
株式会社丹青社
Tanseisha Co.,Ltd.
〒110 東京都台東区上野5-2-2
5-2-2, Ueno, Taito-ku, Tokyo, 110
Tel.03-3836-7494

●
株式会社丹青社オフィスデザイン研究所
Tanseisha Co.,Ltd.Office Design Institute
〒113 東京都文京区白山1-37-6 東信白山ビル
Toshin-Hakusan Buld.1-37-6, Hakusan, Bunkyo-ku, Tokyo, 113
Tel.03-5689-0857

●
株式会社テイ・グラバー
T.Glover Co.,Ltd.
〒141 東京都品川区西五反田3-7-14興和三信ビル9階
9F, Kowa Sanshin Bldg.3-7-14, Nishi-Gotanda, Shinagawa-ku, Tokyo, 141
Tel.03-3779-6121

●
ディエム・デザイン株式会社
DM Design Co.,Ltd.
〒107 東京都港区南青山3-2-6
3-2-6, Minami-Aoyama, Minato-ku, Tokyo, 107
Tel.03-3404-2745

●
株式会社都市計画
Toshikeikaku Inc.
〒564 大阪府吹田市豊津町2-1第2中田ビル4階
4F,#2 Nakata Bldg., 2-1, Toyotsu-cho, Suita-shi, Osaka, 564
Tel.06-338-8800

●
日建設計
Nikken Sekkei Ltd.
〒117 東京都文京区後楽1-4-27
1-4-27, Koraku, Bunkyo-ku, Tokyo, 117
Tel.03-3813-3361

●
日建設計
Nikken Sekkei Ltd.
〒460 愛知県名古屋市中区栄4-15-32
4-15-32, Sakae, Naka-ku, Nagoya-shi, Aichi, 460
Tel.052-261-6131

●

●
日建設計
Nikken Sekkei Ltd.
〒541 大阪府大阪市中央区高麗橋4-6-2
4-6-2, Kouraibashi, Chuo-ku, Osaka-shi, Osaka, 541
Tel.06-203-2658

●
株式会社日展
Nitten Co.,Ltd.
〒110 東京都台東区東上野6-21-6
6-21-6, Higashi-Ueno, Taitou-ku, Tokyo, 110
Tel.03-3847-4111

●
株式会社日本設計
Nihon Sekkei, Inc.
〒163 東京都新宿区西新宿2-1-1
2-1-1, Nishi-Shinjuku, Shinjuku-ku, Tokyo, 163
Tel.03-3344-2311

●
野中建築事務所，野中卓
Nonaka Architect & Associates, Takashi Nonaka
〒862 熊本県熊本市大江4-7-20
4-7-20, Ohe, Kumamoto-shi, Kumamoto
Tel.096-364-3044

●
株式会社乃村工藝社
Nomura Co.,Ltd.
〒064 北海道札幌市中央区大通西9丁目丸菱ビル5階
5F, Marubishi Bldg.,9-Chome, Odori-Nishi, Chuo-ku, Sapporo-shi, Hokkaido, 064
Tel.011-231-3350

●
株式会社乃村工藝社
Nomura Co.,Ltd.
〒108 東京都港区港南3-8-1森永乳業港南ビル
Morinaga-Nyugyo Konan Bldg, 3-8-1, Konan, Minato-ku, Tokyo, 108
Tel.03-5479-2174

●
株式会社乃村工藝社
Nomura Co.,Ltd.
〒108 東京都港区芝浦4-6-4
4-6-4, Shibaura, Minato-ku, Tokyo, 108
Tel.03-3455-1171

●
株式会社乃村工藝社
Nomura Co.,Ltd.
〒556 大阪府大阪市浪速区元町1-2-6
1-2-6, Motomachi, Naniwa-ku, Osaka-shi, Osaka, 556
Tel.06-633-3331

●
株式会社乃村工藝社
Nomura Co.,Ltd.
〒559 大阪府大阪市住之江区東加賀屋1-11-26
1-11-26, Higashi-kagaya, Suminoe-ku, Osaka-shi, Osaka, 559
Tel.06-686-3331

●
株式会社長谷エコーポレーションエンジニアリング事業部
Haseko Corporation Engineering Dept.
〒105 東京都港区芝2-32-1
2-32-1, Siba, Mimato-ku, Tokyo, 105
Tel.03-3456-5473

●
株式会社三橋建築設計事務所
Mihashi Architectural Office
〒991 山形県寒河江市大字寒河江2517
2517, Sagae, Sagae-shi, Yamagata, 991
Tel.0237-84-4358

●
YBP 設計室(野村不動産+大林組)
YBP Design Room
(Nomura Real Estate Development Co.,Ltd.+Obayashi Corporation)
〒240 神奈川県横浜市保土ヶ谷区神戸町134
134, Godo-Machi, Hodogaya-ku, Yokohama-shi, Kanagawa, 240
Tel.045-333-6611

●

後書き

この「環境デザイン ベストセレクション4」の作品は、1991年初頭までに完成したもののうち、特に優秀なものが選ばれて、日本全国から集められた。ご覧のとおりデザインのレベルは、年を追うごとに確実に上がってきている。特に、レジャー、商業施設、公共環境、オフィス環境の作品群には、いま話題の「優れもの」が勢ぞろいした。一方で、ランド、地域開発、ウォーターフロント部門の作品数が比較的少なかったが、次に述べるように、あと数年後には「環境デザイン」が主役になるであろう。1991年には、東京都および近県を含めての「東京圏」に日本の人口の4分の1が集まり、移り住むといわれる。その人口集中の受け皿として、東京湾を取り巻く横浜のMM21計画、千葉県の幕張新都心計画がある。これらが中核となり、新しい日本の都市構想をサポートしている。また、大阪湾でも1995年開港を目指して工事が進められている関西国際空港、その周辺のテクノポート大阪、神戸ポートアイランド2期、六甲アイランドなど、ウォーターフロント開発が目白押しである。

これらの計画がすべてが実現するには、あと5年から10年が必要であろう。しかし、もうすでに一部が実現している。その他には、テクノポリス、テレポート、リサーチパーク、インテリジェント・シティ、リゾート開発、地域イベントなど、数え切れないほどの「環境デザイン」の誕生する可能性がある。
このように大都市を背景にした開発は、大きな変貌期を迎えている。今から1990年代の半ばにかけて、「環境デザイン」のニーズが高まるに違いない。その状況にじっくり注目し、取材を続けなければならない。
環境デザインの生き生きした「いま」を記録するこの本の発刊は、将来のためにも意義深い。ここに選ばれた多くのクリエーターと出品関係者に敬意を表し、感謝を申し上げたい。

「環境デザイン ベストセレクション 4」編集委員会

This book of Environmental Design Best Selection 4 contains the best works completed throughout Japan until early 1991.

As can be seen, the levels of design are increasing steadily year by year. We have concentrated on introducing some excellent work on the present-day talking points of leisure, business concerns, public environs and office environs. On the other hand, work based on the fields of land, area development and waterfronts are comparitively few, but, as explained later, these will become the main talking points of environmental design in the near future.

It is said that one quarter of the population of Japan will be collected within the Tokyo Metropolitan and surrounding prefectures within 1991. The main saucer for this population increase will be the areas surrounding Tokyo bay with the Yokohama MM21 planning, and the Makuhari city planning in Chiba.

Using these areas as the main objectives, support for the concept of new Japanese cities will evolve. Additionally, there is much waterfront construction expected to be completed by 1995 in the Osaka bay region which includes the Osaka International Airport, the nearby Technoport Osaka, the 2nd part of development for the Kobe Port Island, the Rokko Island, etc.

I will take another 5 to 10 years to completely actalize all of these plans. However, certain parts are already complete.

Furthermore, there is also the possibility of seeing the birth of futher environmental design projects in the style of technopolises, teleports, research parks, intelligent cities, resort development and regional events. In this way, development which has large cities as its base is greatly changing. There is no doubt that the need for environmental design between now and the mid-1990's is on the increase. We therefore need to pay much attention to the situation and continue the required coverage. The publication of this book which records the present situation of environment design will have a deep effect on the future. I would like to take this opportunity to give my gratitude to all of the creators who were selected for this publication and all others who contributed their work.

Environmental Design Best Selection 4 ： Editorial committee

環境デザイン ベストセレクション 4
Environmental Design Best Selection 4

発行　1991年6月25日初版第1刷発行

編集　グラフィック社編集部
監修　オレンジブック
　　　寺澤勉
装丁　株式会社高間デザイン室
本文デザイン　株式会社イズ：清野尹良
翻訳　株式会社新世社
　　　株式会社ウイン
編集協力　杉本正顕
　　　近藤繁
　　　森林兵衛
　　　近宮健一
　　　水野晃一
　　　松川秀樹
　　　池上俊郎
　　　藤田鉱太郎
　　　斉藤明男
発行者　久世利郎
発行所　株式会社グラフィック社
　　　〒102 東京都千代田区九段北1-9-12
　　　Tel.03-3263-4318 振替・東京3-114345
印刷・製本　錦明印刷株式会社
写植　三和写真工芸株式会社

定価13,000円(本体12,621円)